THE . S

THE POWER OF DESERTS

Climate Change, the Middle East, and the Promise of a Post-Oil Era

DAN RABINOWITZ

stanford briefs
An Imprint of Stanford University Press
Stanford, California

Stanford University Press
Stanford, California

Printed in the United States of America
on acid-free, archival-quality paper

Cataloging-in-Publication Data available on request.
ISBN: 9-781-5036-0998-3 (paper)

Cover design: Rob Ehle

Cover: Acacia tree in the Ein Khadra oasis, Egypt. Walaa | Wikimedia Commons; Solar array background pattern, Rob Ehle

Typeset by Classic Typography in 11/15 Adobe Garamond

CONTENTS

THE POWER OF DESERTS

INTRODUCTION

The Heat Is On

The towering granite mountains of South Sinai, home to the Jebaliya Bedouin tribe, also host the Greek Orthodox monastery of Santa Katerina, built in the sixth century by Byzantine emperor Justinian. A Jebaliya legend recounts a drought centuries ago, so long and bitter that the tribe's survival was at stake. Desperate, the elders sought help in the monastery. The council of monks heard their plight, prayed, and came up with a remedy: a parchment scribbled with some foreign text. "Take this scroll," the bishop told a Bedouin youngster, "climb Mount Sinai, and place it at the top so it is visible from the heavens." Turning to the elders, the clergy then proclaimed: "God willing, the sacred verse will open up the heavens. Soon water will pour down the mountains, filling wells, reviving pastures."

The boy clutched the parchment and embarked on the ascent, but halfway up the mountain he had a change of heart. Instead of making for the summit, he headed towards a creek where his family had a home and pastures and placed the precious scroll there. He then descended to the monastery to inform the others that his mission was accomplished, keeping the exact location where he put the scroll a secret.

Heavy rains did pour that night, but only on the ravine in which the parchment lay. Too small to hold the gushing torrent, the gorge soon flooded, sweeping the young man's family, their home, and flock to horrid death. Forty days of mourning followed, after which the tribal elders made a somber vow: never again would any member of the Jebaliya tribe defy an edict issued by the monks.

Acknowledgment of human vulnerability to nature's wrath is a familiar trope in Middle Eastern lore. The Sumerian odyssey of Gilgamesh, written almost 5,000 years ago, has Utnapishtim saving kin, livestock, and plants by sailing in a reed boat to escape a flood, a feat for which he earned eternal life. The tale of the great deluge recorded in the Old Testament two millennia later likewise had Noah loading animals and plants onto an ark to avert calamity. Jacob's dozen sons escaped famine-stricken Kena'an to look for sustenance in Egypt. The youngest, Joseph, later rose to prominence in Pharaoh's court by accurately predicting seven lean years and by advising Pharaoh to stock up on grain, which saved the kingdom.

These legends underscore prescient leadership. Utnapishtim and Noah learned of the impending floods through divine revelation; Old Jacob realized early enough that a drought was coming Kena'an's way and that salvation for his offspring was awaiting on the Nile; Joseph's premonitions came in dreams. But prescience was not enough: all the protagonists in those ancient myths were persuasive leaders too, convincing their peers that danger was approaching and successfully mobilizing their respective communities for timely, costly efforts to attain deliverance. The Jebaliya legend adds solidarity as an essential value: to avoid disaster, it suggests, people must put their collective interests above narrow individual considerations.

The Middle East has seen numerous climate fluctuations in the Common Era, some of which left evidence behind. A warm and dry period in the first century was reflected in textual records and material relics.[1] A warm spell in the sixth century is believed to have expedited the fall of the Roman empire,[2] and a climatic shift early in the second millennium decimated Europe's agricultural output, pushing bankrupt peasants to become crusaders and colonize the Holy Land.[3] A dry spell that hit the region in the 14th century, concurrently with Europe's Little Ice Age, forced villagers to abandon sedentary agriculture and become nomadic pastoralists,[4] a migration which transformed the Muslim world, energized the Ottoman expansion westwards, and sent shock waves across Europe.[5] In the 17th century a long dry spell in southeast Turkey brought the Ottoman Empire to the brink of implosion.[6]

Unlike earlier episodes, contemporary global warming is anthropogenic. The Industrial Revolution created a formidable dependence on fossil fuels for manufacturing, locomotion, and power generation. As *Homo sapiens* morphed into *Homo sapiens combustans,*[7] vast amounts of CO_2 emitted from burning fossil fuels intensified the atmosphere's greenhouse effect, gradually warming the planet.[8] And while earlier climate events crept up on humans unannounced, contemporary global warming has been signaling its imminent arrival since at least the 1980s,[9] opening a narrow window of opportunity in which remedial action could perhaps still prevent our climate from spinning out of control.

A community's ability to withstand the impending post-normal climate era will hinge on how it does on three related counts: impact, adaptation, and vulnerability.[10] *Impact* denotes the extent to which pertinent biospheric conditions are disrupted. Will the temperature range required for important food crops be sustained? Will the natural water cycle hold? Are coastlines going to remain where they currently are? Can biological diversity stay intact? *Adaptation* is the human capacity to mobilize economic, technological, and social assets in an effort to come to terms with detrimental impact. *Vulnerability* pits response against potential impact to yield projected levels of exposure.

The outlook for climate impact, adaptation, and vulnerability in the Middle East and North Africa is, to put it mildly, disconcerting.[11] Much of the region, stretching south to north from Yemen to Turkey and east to west from Iran to Morocco, falls within the Northern Hemisphere's desert belt, one of the hottest, driest regions of the world.[12] Predictions suggest that by the end of the century average summer temperatures across the region could be 4°C above their current levels.[13] Areas in the northern Middle East, where populations are denser and agriculture more developed, could see rainfall dropping 25 percent before the century is out.[14]

Climatic shifts on such a scale can inflict a blow on any territory. In a region as arid as the Middle East, where even minor fluctuations carry drastic consequences, they could be devastating. Areas which currently sustain seasonal pasture and rain-fed agriculture could turn to wastelands. Water cycles, stretched as they are, could be depleted further. Chronic heat stress could render major cities unfit for human life.[15] Sea-level rise could submerge ports and other seaside installations, salinize coastal aquifers, and force millions to migrate.[16]

The economic rift between the richer countries of the Middle East and their poorer neighbors is already creating an adaptation divide. As climate stress intensifies, richer countries in the region will rely on their accumulated wealth and technological prowess. Poorer nations, on the other hand, could meet rapid decline and suffer from devastating chain reactions whereby climate change creates scarcity, insecurity, instability, and forced migration. Qatar, United Arab Emirates, Oman, Abu Dhabi, Bahrain, Kuwait, and Saudi Arabia boast GDP per capita akin to levels found among the world's richest nations. Along with Israel and some metropolitan areas of Turkey,

Iran, Lebanon, and Morocco, their civil infrastructures could, with luck, provide protection even in the post-normal climate era. Meanwhile, poorer countries such as Yemen, Syria, Sudan, Somalia, Jordan, or Egypt, where large communities rely on subsidized staples for basic sustenance and lean on fragile infrastructures, will be infinitely more exposed to climate tremors.

Toxic blends of economic strife, inequality, and socio-economic tensions can send entire societies into spirals of dysfunction, distrust, and despair.[17] Isis and Al-Qaeda rose to prominence in Iraq and Syria by claiming to replace governments that abandon marginal communities. The Israeli-Palestinian conflict, tensions between Iran and Sunni regimes, rent-seeking interventionism on the part of regional and global powers, and other tensions could further exacerbate climate-related vulnerabilities in the region.

Communities across the Middle East are already experiencing hardships partly attributable to climate change.[18] Brutally long dry spells in Syria and Sudan, combined with other circumstances, forced millions to abandon hinterlands no longer tenable for farming. Seeking better lives in urban hubs, many of those displaced find to their disappointment that governments, already struggling to cater to the urban poor, are reluctant or unable to help them settle in their new locations.[19]

The situation in the Middle East is further complicated due to the historic role the region played in the political economy of oil. The region's oil exporting countries, which among them account for more than 40 percent of global oil production[20] and control more than half of known global reserves,[21] have oil accounting for the lion's share of their GDP. Like oil producers elsewhere, they tend to see efforts to curb climate change, which invariably entail downsizing oil production, with suspicion. Poorer Middle Eastern countries on the other hand, whose chances to withstand the vagaries of climate change are slim, pin hopes on a comprehensive global pact to save them. A cruel schism thus emerges between humbler communities who yearn for climate action, and other more economically robust ones who often drag their feet. This tension, between countries and sometimes between particular populations in them, is an important focus of attention within this volume.

Climate change is more than yet another environmental hiccup. Systemic, comprehensive, and extremely potent, this crisis of our era is already recasting knowledge and reshaping scientific methods. Atmospheric science grew from a secondary subfield in geophysics to a burgeoning powerhouse attracting talent and investment, and yielding important insights. A new discipline, Earth System Science, has emerged.[22] Glaciology (the study of glaciers), dendrochronology (inferring past climate patterns from tree rings), volcanology, paleozoology, and other branches of the natural sciences that regularly produce insights applicable to climate modeling have grown substantially.[23] Much research in biology, agriculture, and forestry now focuses on enhancing the productivity and endurance of species capable of surviving the post-normal climate era. Research in new energy forms and water technologies, and of policies that could enhance efficient climate regulation, attract more innovative power than ever before.

Scholars in the social sciences and the humanities too have made important contributions to the conceptualization of climate change and the analysis of its attendant perils. Historian Dipesh Chakrabarty suggested climate change is formidable enough to merit a departure from conventional historiosophical conventions.[24] Local, national, and regional histories, he argues, cannot capture the significance and scope of current climate shifts. The time has come for them to be eclipsed by a "species history," where humanity at large becomes the object of inquiry.

Sociologist Ulrich Beck made a forceful argument about the challenges the climate crisis poses for social science theory.[25] Global warming, he noted, is already transforming patterns of political accountability, compelling theorists to forge new vocabularies. Bruno Latour's comments on the return of nature into contemporary life point in the same direction.[26] Early modernism, he argues, promised to undo uncertainties that emanated from nature's untamed forces. As dams diverted rivers and levees elevated shorelines, as mines leveled mountains and cities kept expanding to the countryside, nature was relegated to quaint reserves, becoming a nostalgic fantasy. Then with climate change the wrath of nature was unleashed anew, reintroducing ancient planetary connections and courting potential disasters on an unimaginable scale. The problem, according to Latour, is that the need for collective universal action confronts the international community at a time in which it has no globe to work with.

Ideally, the social sciences should provide conceptual tools to tackle any crisis. Well not in this case, says Latour. For almost two centuries, social scientists have focused on the state as the primary, self-evident unit of analysis of structures, relations, and connections. Now when they are called to theorize collective action on a global scale, social scientists are stuck, hemmed in by the philosophical and analytical cul-de-sac created by the state.

These calls to recast sociology, politics, and historiography for better comprehension of the climate crisis are fundamentally important. They also signal an epistemological reversal which defines current environmental thinking. The logic that guided 20th-century environmental problem-solving looked to other scientific disciplines for explanatory theory, methodologies, and practical solutions. The systemic nature of climate change and its immense disruptive power now force scientists to reexamine their own knowledge fields. Instead of scientific disciplines propelling environmental praxis, we now have a looming biospheric crisis compelling scientists to reflect, reexamine, and reinvent their disciplines.

This reversal is highly relevant for regional analysis too. Understanding the Middle East and predicting its future without taking into account the impact of climate change is no longer tenable.

In September 2019, 16-year-old Greta Thunberg made a memorable appearance at the UN General Assembly in New York. Her mesmerizing speech, delivered in front of dozens of heads of states and watched by millions globally, embodied the moral authority now claimed by youngsters as they demand that older generations do what it takes to avoid climate chaos. Solemnly resolved and unflinchingly direct, Thunberg conveyed a sense of urgency unlike anything seen on stages of the international climate forums before.

To prevail, the logic personified by Thunberg goes, environmentalists in Western liberal democracies must begin by acting as envoys of scientific evidence regarding global warming. The truth they speak will then help them win over the mainstream, give them political influence, and eventually enable them to strong-arm governments into forcing corporations and consumers to dramatically reduce emissions.

For this mission, Greta is indeed the most effective emissary. She knows the science and political economy of the climate crisis inside out and provides the necessary facts and figures with inspirational conviction, clarity, and poise. She has a blueprint for immediate action that is coherent, sensible, and practical. She talks the talk and walks the walk and finds her way to hearts of folks.

But as the emotional impact of her stirring performance at the UN subsides, the formidable obstacles that await those who seek to realize her vision bounce back. Will the wave of consciousness Greta so movingly created last long enough to revolutionize the global energy market? Can her ideas impact realpolitik and overcome the self-perpetuating logic of the global capitalist economy? Can the alternative she propagates materialize before the fleeting window of opportunity we now have—post-denial, pre–final despair—dissipates? Will the strategy she now personifies win our fateful race against time?

Many, myself included, hope the answer to these questions is affirmative. That said, this book's analysis of climate change and energy in the Middle East draws a different roadmap to climate redemption. What if instead of the planet being saved by highly educated, conscientious young environmentalists in liberal democracies, it is redeemed by a tiny group of Middle Eastern oligarchs attempting to advance their own self-interests? What if instead of lofty principles of sustainability and climate justice, the sensibility that eventually drives the renewable energy transition comes from despots seeking to protect their hypermodern, ridiculously expensive lifestyles? Could the long-awaited green revolution begin as a result of attempts on the part of the current lords of global energy to prolong and further tighten their control?

To explore this counterintuitive trajectory, this book looks at an assemblage of challenges and opportunities the region faces as the post-normal climate era looms. Chapter 1 reviews the region's somber climate future. Chapter 2 shows how climate inequality, between states and, not least, within them, exacerbates the crisis. Chapter 3 looks at the disruptions climate change could bring to regional security and demographic stability, and at the unattractive prospect of climate refugees. Chapter 4 explores the potential of solar energy as an economic game changer, focusing on the dilemmas this avenue poses for the region's oil producers. Chapter 5 then comes full circle to the notion that the Middle East could spearhead the transition to renewables, analyzing the circumstances and dynamics which might help this counterintuitive option become a reality.

1 PARCHED FUTURE

In June 2018 the Iraqi Ministry of Agriculture imposed a countrywide ban on rice cultivation for the coming summer.[1] Rice has been a staple of the local diet since pre-Islamic times,[2] but the drastic measure by the government seemed inevitable. A brutal spell of droughts which hit the region in preceding years decimated flows in the Euphrates and the Tigris, the twin rivers which between them supply 98 percent of Iraq's water.[3] Residents of Baghdad waded knee-deep across the Tigris that spring. In Kurdistan dead fish piled on banks of dried-up tributaries.[4]

A water crisis had been brewing in neighboring Iran too. In 2014 the UN Development Program predicted that by 2025 Iran's available water supply per capita would be half what it had been in 1990.[5] In June 2018 urbanites and farmers took to the streets of Khorramshahr and Ibadan to protest against the chronic water shortage, claiming the situation was exacerbated by mismanagement and favoritism on the part of government.

Four people died from police fire, triggering more rallies and clashes in the coming days in additional Iranian cities.[6]

Both Iraq and Iran boast ancient agricultural civilizations where agro-administrative institutions have regulated water and food production since the Bronze Age. In medieval times, when Europe was a sparsely populated, unproductive jumble of feudal fiefdoms, Baghdad was the capital of the Abbasid Caliphate and the epicenter of the prosperous Islamic Golden Age. This history of the eastern edge of the famous Fertile Crescent makes the water crisis of 2018 all the more poignant. Anything but trivial, such extraordinary mishaps could soon become the new normal.

A recent forecast of temperature change across the Middle East in coming decades used a combination of heat and humidity known as wet-bulb temperature (WBT).[7] Thirty-five degrees WBT is considered a threshold above which even able-bodied humans cannot safely spend any significant amount of time outdoors.[8] The study indexed a number of important cities along the shore of the Arabian Gulf, including Doha in Qatar, Dubai and Abu Dhabi in the United Arab Emirates, Daharan in Saudi Arabia, and Bandar e Mahshahr in Iran as likely to regularly hit the 35 WBT mark by 2060, thus becoming unfit for human habitation. Dubai and Abu Dhabi, two of the most extravagant and affluent metropolitan centers globally, could hit this threshold even if climate change proves relatively mild.[9] As heat stress mounts, labor conditions, business routines, and behavioral norms will need to be adjusted, potentially radicalizing occupational stratification further. Households with young, old, or infirm members will seek to relocate if they can afford to. Those with lesser means will stay behind, enduring ever worsening conditions.

Predictions for cooler places in the region suggest they too could face heat-stress challenges. Summer temperatures in Beer-Sheva, a mid-sized town in the Northern Negev desert in Israel, currently render outdoor activities in July, August, and September between 11 a.m. and 3 p.m. impossible. By 2060, heat stress in the town could extend from 9 a.m. to 5 p.m. and last from June to early October.[10] The disconcerting vision of modern cities turning to temporary or even permanent ghost towns due to direct heat stress on the human body is, of course, unsettling. But direct damage to the human body is not the primary toll climate change is likely to exert. Failed crops, dwindling water supplies, rising sea levels, storms, and diseases due to wider distribution of insects are much more pertinent threats.

Climate models use the past as a baseline, then try to isolate those factors that might force future trajectories to depart from the expected pattern. Relatively easy to construct, these models work like architectural or engineering mock-ups, allowing users to add new features, tweak estimates and forecasts, test probable outcomes, and modify solutions.

Climate modeling is simple when the climate remains stable. It is manageable even at times of change, provided shifts are linear and constant.[11] Contemporary global warming, alas, is neither. While future amounts of greenhouse gases discharged into the atmosphere are fairly easy to estimate, calculating the impacts they might have on global temperatures is more complex. Forever comparing their past estimates with real-time data, climate scientists often find it difficult to determine the pace of things to come. The post-normal climate era, in short, defies precise prediction.

A first attempt at a standard global framework for climate predictions came in 1995, with the Second Assessment Report of the Intergovernmental Panel on Climate Change (IPCC)[12]—the scientific forum established to provide the UN's Framework Convention on Climate Change (UNFCCC) with updated, comprehensive climate data. Looking to the 21st century, the report introduced four scenarios or "storylines" of global economic, demographic, and technological trends.[13] Each storyline involves a different estimate of future greenhouse gas emissions and, concomitantly, a different forecast for future climate shifts. By 2000, the four storylines had been developed into 40 subscenarios, detailed in a document whose title gave the methodology[14] its name: Special Report on Emissions Scenarios (SRES).[15] The science-based, quantified predictions SRES provided were too detailed for the popular media to carry and too nuanced for the pubic to digest. A pioneering yardstick, it was embraced, however, by scientists and planners grappling with a complex and confusing future. The predictions premised on it prompted a sense of urgency, alerting politicians and administrators to the problem. This methodology informed climate assessments for over a decade, facilitating early attempts at scaled-down models that informed regional and local forecasts.

The ability to take a global climate model and scale it down is especially important in climatically vulnerable regions. The first models pertaining to the Middle East appeared in the late 1990s. A decade later, a much cited, comprehensive study of the region's future climate premised on the SRES's "business as usual" (A2B) storyline made three principal predictions.[16]

First, temperature hikes in the region will be significantly steeper than those predicted for the world at large. By 2050, average yearly temperatures across the Middle East could be 1.5 degrees centigrade higher than they were in 2005; by 2100 they could be 4 degrees above their 2005 figure—double the projected planetary rise. Temperatures in coastal areas will climb more moderately than inland, except along the Arabian Gulf coast. Already hotter than most other parts of the Middle East, by 2050 the Gulf could see temperatures 4.2 degrees higher than today.

Two additional forecasts included in the study pertain to rainfall. First, by 2050 the Middle East, already arid, could see average precipitation decrease by 3 to 4 percent, a drop that could reach 10 percent by century's end.[17] The second forecast indicates differences between the northern part of the region and the southern. In the south, proximity to a warming Indian Ocean could actually trigger an *increase* in precipitation of up to 25 percent.[18] The north, in contrast, will see a steep decline in rainfall.

Significantly, declining rainfall in the north, an area which includes Syria, Lebanon, Israel, Turkey, northern Iraq, and northern Iran, would be more consequential than added rainfall in the south. Communities in the north, we should remember, rely on rain-fed agriculture, and a 25 percent drop in rainfall there could push the annual figure in many parts below 200 millimeters,[19] the all-important threshold for rain-fed cereal crops. Also, losing a quarter of the annual precipitation falling on the upper drainage basins of the Tigris and Euphrates, which translates to about 125 millimeters,[20] would deplete water volumes downstream, crippling food crops and economic viability in eastern Turkey, northeast Syria, and Iraq.

In the south, in contrast, where current rainfall quantities are considerably smaller, even a 25 percent increase will only bring the yearly average to under 100 millimeters, a far cry from the 200 millimeters needed for rain-fed crops. The study's bottom line is startling: by 2100, it predicts, some 170,000 square kilometers in the Middle East which currently sustain agriculture will no longer be arable. It also predicts that by mid-century, dry seasons will be two to four weeks longer than they are now, a figure likely to expand to ten weeks by 2100.[21] The prolongation of what is already an unbearably long, hot, and dry season will have devastating consequences for sheep, goat, and camel pastoralism, sectors that are normally considered more resilient to environmental fluctuations.

A climate model for the Mediterranean zone provides a similarly bleak prognosis.[22] Using simulations based on three different storylines, it anticipates a northward shift of the Atlantic storm track that will result in a significant reduction of rainfall in spring and summer. Combined with robust warming and a higher frequency of extreme heat events, the Mediterranean too is looking at an unattractive climate future.

In 2010 the limitations of the SRES methodology[23] pushed scientists to come up with a new method labeled representative concentration pathways (RCP).[24] Instead of extrapolating future amounts of greenhouse gas emissions from complicated projections of global demographic growth and economic activity, the new method uses a more direct approach. Atmospheric warmth emanates from the Earth's surface, where sunlight is converted to heat energy, and is determined by the amount of greenhouse gases present to retain that heat. The difference between the amount of energy currently retained and the amount that might be trapped in the future, when greenhouse concentrations rise, is calculated and embodied in a simple figure.[25] Known in scientific circles as radiative forcing, this X factor of global warming serves as a reliable index of future fluctuations.[26] Using it to guide predictions is simpler. Predictions no longer hinge on a complex set of rigid forecasts involving multiple indicators, and model designers can update their estimates as real-life observations become available.

Scientists using RCP normally refer to four scenarios, each characterized by a different quantity of extra energy likely to be trapped in the atmosphere in the future. The most pessimistic of the four scenarios, labeled RCP 8.5, assumes a relatively high amount of extra energy trapped as heat in the atmosphere.[27] Two intermediate scenarios are known as RCP 6.5 and RCP 4. The fourth, most optimistic scenario, known as RCP 2.6, assumes the smallest amount of extra heat retention by the atmosphere.[28] RCP's flexibility also enables scientists to adjust global models to local contexts more easily.[29] It informed the Fifth Assessment Report of the Intergovernmental Panel on Climate Change of 2013,[30] which guides most ongoing global and regional climate predictions and will be an integral component in the panel's Sixth Assessment Report due out in 2021.[31]

The Atlas of Global and Regional Climate Projections[32] uses the relatively mild scenario RCP 4.5 to develop an array of climate predictions, including ones which are specific to the Middle East.[33] The section on the region begins with a prediction for significant increases in both summer and winter temperatures. By 2035, it suggests, temperatures across the region will be 1.25 degrees above their 1990s level,[34] a hike likely to reach 2.5 degrees by 2065 and 3.5 degrees by 2100. Coastal stretches along the Mediterranean and the Indian Ocean will probably fare marginally better, with temperatures in 2100 only 2.5 degrees above their 1990s level.

Second, the atlas provides forecasts of future rainfall levels, a more complicated task but not more reassuring. By mid-century, it suggests, parts of the Middle East that are nearer to the Indian Ocean in the south could see precipitation in the rainy season (October to March) increase by as much as 45 percent compared to amounts in the 1990s. Summer rains will surge by 30 percent as well.[35] The rule of thumb in this part of the region seems to be that steeper temperature increases will translate to more significant rise in rainfall.

In arid zones, however, where the intensity and frequency of rain events are unpredictable, annual averages provide only a partial picture. An erratic regime of rainfall is bad for water cycles, agriculture, and pastoral production. A forecast of future droughts in the Middle East published recently by the American Meteorological Society suggests in fact that the eastern part of the Mediterranean is likely to become much drier.[36] This forecast further indicates that stronger and more frequent droughts will cause vegetative growth to shrink, which will in turn increase the vegetation albedo effect. The bottom line is a problematic feedback loop whereby reduced vegetation cover absorbs less energy, causing more of it to be reflected to the atmosphere as heat, thus further accelerating desertification.[37]

These advances in prediction methodologies do not merely provide more accurate regional models. Climate scientists can now produce more nuanced forecasts for individual states, districts, counties, and even cities.

A recent model suggests that Saudi Arabia's average daily temperature highs in summer in the year 2090 will be 37.5°C—5 degrees above its 1990s level.[38] And while overall rainfall is likely to go up a little,[39] its blessings will be curtailed by a detrimental seasonal drift. Rains, which currently come mainly in the spring, could be delayed to early summer, when overheating is likely to arrest the maturation of rain-fed crops and pastures. For Saudi Arabia, which used the wealth created in the oil boom to substantially increase its agriculture sector since the 1980s, such seasonal shifts would be detrimental.

Projections for the future of water in Saudi Arabia suggest that industrial water demand, already on the rise, could surge by an additional 50 percent by 2030.[40] This will exert more pressure on existing water sources, further taxing agriculture. Malnourishment, which has been virtually absent from the country for almost a century, could be back. One estimate in fact suggests that by 2100, food shortages could impact up to a third of the kingdom's population.[41]

The United Arab Emirates did not invest as much as Saudi Arabia in agriculture. As a result, the preoccupation on the part of the Emiratis with climate change focuses primarily on sea-level rise, more severe sand storms, and the intensification of urban heat islands.

The Intergovernmental Panel on Climate Change predicts that sea levels across the planet will rise by almost a meter this century. Forecasts for the UAE premised on this figure suggest that by 2100 the country could lose as much as 5,000 square kilometers to the surging tide, a figure representing 6 percent of its entire landmass.[42] With much of the population concentrated along the Arabian Gulf, many will be forced to relocate. UAE government sources indicate that by 2100 the sea surge could impact 85 percent of the population and 90 percent of the country's infrastructure.[43]

A study analyzing sand and dust storms suggests that a recent increase in the frequency and strength of windy outbursts in the region is consistent with the advent of climate change.[44] It identifies shifts in wind patterns over the Gulf as a contributing factor to the further expansion of the country's arid landscapes.

With rainfall providing less than 1 percent of the UAE's water demand, arable land is in immensely short supply. Agriculture in fact accounts for only a marginal fraction of the country's gross domestic product,[45] and food is mostly imported. Significantly, the UAE's main suppliers—India, Sudan, Morocco, and Pakistan—are themselves susceptible to yield shocks due to global warming.[46] Willingness on their part to supply crops to the UAE indefinitely, perhaps at the expense of their own populations, cannot be guaranteed.

A study that assessed potential impacts of climate change on Israel, which assumed accelerated global economic and demographic growth, came up with alarming forecasts.[47] By 2100 summer temperatures in Israel could be

3.3 degrees centigrade higher than their current level, a significantly steeper rise than the 2.8 degrees the same development scenario predicts for the world at large. Warming, the study indicates, will affect nocturnal temperatures more than daytime conditions, making cooler evenings less frequent. Heat waves will be longer and more frequent, and rainfall could drop 30 percent. Snow in Galilee and in Jerusalem, already rare events, will become almost nonexistent. A more recent forecast for Israel premised on a milder climate change scenario similarly predicts that by mid-century temperatures will be 2.5 centigrade higher than they were in 2000, with precipitation falling by as much as 40 percent.[48]

In the second half of the 20th century, average Israeli winters saw three prolonged dry spells (eight days or more with no rain).[49] This figure could double by 2100. Shorter dry spells (three consecutive days with no rain) could likewise leap from seven per winter in the 20th century to double that amount in the late 21st. This while the number of wet spells (three or more consecutive wet days) could drop from 20 to approximately 10 per winter.[50]

Predictions for the frequency and length of future droughts present a diametrically opposed count. A drought spell is defined as two or more years in which precipitation falls 30 percent or more below average. A model developed for the Eastern Mediterranean, an area covering Cyprus, Israel, Lebanon, southern Turkey, and parts of Syria, suggest that whereas the 20th century saw 12 such spells, the 21st century could witness up to 20. Future droughts in the region are likely to be longer, some lasting up to seven years, with serious consequences for the natural water cycle.

In Israel, water shortages associated with a typical drought spell in the 20th century were equivalent to the amount of water added to the natural water cycle by 1,000 millimeters (40 inches) of rain. A typical drought spell in the 21st century, expected to be longer and more intense than those known in the 20th century, could create water deficits that will be three times bigger. Average annual flow into the Galilee basin, Israel's primary water reservoir, could drop from around 375 million cubic meters (m^3) per annum in 2000 to 305 million m^3 per annum in the future. The drought line, delineating areas where annual rainfall fails to reach the 200 millimeters threshold, will nudge northward, pushing tracts which currently sustain rain-fed agriculture to become marginal. Rain will come in fewer, more violent events, with larger quantities washed away as floods into the sea. The quality of water flowing to Lake Galilee and to recently constructed artificial reservoirs will be impaired by higher nutrient concentrations due to intensified soil erosion.

Tree rings provide clues to past climate. Counting them from the outside of the trunk inward can help infer temperature levels and amounts of rainfall in particular past years. Researchers using this method in various locations in the Mediterranean basin, who worked on trees or wood that go back to medieval times, suggest that 1998 to 2012 was probably "the driest 15-year period in the region since the twelfth century", and almost certainly the driest since 1500 C.E.[51]

An exceptionally long dry spell had devastating consequences for Syria's eastern provinces in recent years. Between 2006 and 2009 Syrian wheat yields fell by half and barley by two-thirds, decimating livestock numbers by a staggering 85 percent.[52] In 2008 Syria's minister of agriculture communicated to the UN that the economic fallout from the drought was "beyond our capacity as a country to deal with."[53] In 2011 and 2012 the UN's Food and Agriculture Organization found that poverty and food insecurity in Syria were so bad that three million people were in urgent need of assistance.[54]

Syria's climatic future could prove to be even bleaker than its recent past. A study by the World Bank indicates that by 2100, September temperatures in central Syria could be almost six degrees higher than their average in 1990, with predictions for other summer months only a fraction milder.[55] Precipitation could grow sparser and more erratic; rain events will be more isolated and intense; drought spells will become more frequent and intense.[56]

Kermanshah, a town situated some 500 kilometers southwest of Tehran at 1,350 meters altitude, provides a representative case of Iran's elevated steppe climate conditions. An Iranian team of climatologists recently developed a scaled-down model for the town's future climate.[57] Their data suggest that the daily minimum temperature in summer, which averaged 14°C in 1980, could climb to 18°C by 2080. Daily maximums in summer, which averaged 36°C in 1980, could rise to 40°C in 2180; maximums in January could likewise leap from 9°C in 1980 to 13°C in 2080.

This Kermanshah project relied on a "business as usual" (A2B) projection for global population, production, and emission.[58] A more recent prediction for Iran, which assumed a significantly harsher warming scenario of RCP 8.5, suggests a leap of summer temperatures by 2080 of almost 6 degrees centigrade compared with figures for the 1990s.[59] The same report also found that temperatures will reach the threshold necessary for plant growth earlier in spring and will remain above that threshold well into the autumn, prolonging the active growth season by 25 days.[60] Many agriculturally significant species have evolved over time by striking a delicate balance between growth and the number of daylight hours, temperature, and moisture in the ground. A longer growing season could throw off this synchronization, causing significant loss in yield.

Half the landmass of Turkey, a rapidly reemerging regional and global power, is cultivated, which is double the ratio in most European countries. With agriculture still a pillar of its national economy, Turkey's fate in the post-normal climate era raises concerns. By 2100, one study shows, the daily maximum temperature in summer could average 7.5 degrees centigrade above its level in 2000, with winter temperatures leaping more than 4 degrees.[61] Turkey's longest dry spell, in other words the longest stretch of consecutive dry days in an average year, could climb from

53 days now to 85 days by the end of the century. Such a leap will drastically reduce the amount of moisture in the ground, arresting a variety of vital biological processes. The active growth season, already 16 days longer than it was in the 1990s, could be extended by another month, with parched crops becoming the new normal.[62]

Climate change, another recent study focusing on Turkey indicates, may not have a massive impact on agriculture until the late 2030s, but will kick in violently then or later.[63] With precipitation likely to decline by 25 percent, regions that currently rely on rain-fed agriculture will lose considerable crop yields. This will in turn catapult food prices up, push poorer populations to the brink of starvation and injure the country's foreign trade.

Egypt, the most populated Arab country, has large segments of the population living below or near the poverty line. Known as "Gift of the Nile," it has climate-related perils lurking on both ends of the great river. At the upper end, disputes over water shares already sour Egypt's relations with upstream riparians Ethiopia and Sudan.[64] Downstream, this extraordinary waterway, which traverses more countries than any other river in Africa, could have its volume drastically depleted due to intensified evaporation and accelerated pumping in the post-normal climate era. Water levels in fact already drop so low in summer that some channels in the lower estuary fail to reach the sea.

The Nile Delta is particularly sensitive to a changing climate. Seventeen thousand square kilometers in size, its isosceles triangle—240 kilometers along the sea and 160 kilometers from north to south—served as Egypt's breadbasket for millennia. With high temperatures and abundant water, it is most suitable for subtropical agriculture, with rice, maize, sugarcane, clover, citrus, and tropical fruit crops taking center stage. Occupying only 2 percent of Egypt's landmass, it accounts for two-thirds of the country's arable land,[65] is home to more than 40 percent of the population, and makes a crucially important component of Egypt's economy.

The delta's 20-meters-thick topsoil, fed by sediments and nutrients transported downstream by the Nile, is also under threat. The construction of the High Dam of Aswan in the 1960s reduced sedimentation to the extent that poorer soils now need intense fertilization to maintain crop yields. As the level of the Mediterranean rises, the power of the Nile to transport particles will grow weaker, challenging the river's capability to compensate for the erosion of the delta's coastal edges.

Almost a fifth of the delta's landmass is lower than sea level; an eighth of it lies between zero and one meter elevation; another eighth between one and two meters.[66] With the Mediterranean already nibbling at the weaker segments of the outer sand belt, should sea-level rise reach two meters, Egypt could lose 45 percent of its productive soil.[67] Damage to agriculture will probably appear long before fields are submerged, as the encroaching sea salinizes coastal aquifers, wells, channels, and lagoons.

A third of Egypt's fish catch comes from sweetwater coastal lagoons on the edges of the delta, now threatened by the rising tide.[68] Al-Ma'adiya, a fishing village east of Alexandria, is situated near the estuary of one of the Nile's lower branches. Elder villagers remember how, as children, they used to walk half an hour from their homes along the channel bank to reach the coast. Now their journey takes ten minutes. Meanwhile, the distance from the mouth of the river, where they moor their fishing vessels, to the open sea has more than doubled. The trip to deeper waters that they used to make in rowboats now requires engine-powered crafts.[69]

Rising sea level also impacts urban neighborhoods. In Alexandria, home to the famous 20-kilometer-long waterfront promenade Al-Corniche, cars have been washed away by sea waves lately, forcing the government to erect concrete coastal barriers.[70] Some beaches, once indispensable components of Alexandria's landscape and sociocultural heritage, have almost vanished recently, much of their sand beds long lost to the rising sea.[71] Historic buildings on the Corniche are crumbling as saline splashes hit bricks, concrete, and woodwork more often and with more intensity than ever. Even if the sea level rises only 70 centimeters, entire neighborhoods will be significantly affected. No wonder Egypt is now listed among the countries most likely to be damaged by climate-related sea-level rise.[72]

Morocco has 3,500 kilometers of coastline along the Atlantic and the Mediterranean, impressive mountain ranges, a significant plateau in the east, and the Sahara desert to the south. With wheat, barley, maize, olives, and sugarcane the staple crops, agriculture is a pillar of the national economy, a bastion of employment, and the backbone of the country's food security.

Research based on a "business as usual" global scenario suggests that in Morocco, rain-fed wheat, barley, and olives are the crops most likely to sustain yield shocks due to climate change. Wheat is projected to decline by up to 26 percent, barley by up to 17 percent, and olives by up to 20 percent. And while outlooks for citrus and for other fruits are mixed,[73] one study predicts that by mid-century climate change could reduce wheat, coarse grains, and sugar, which between them represent half of Morocco's agricultural output and use half the country's arable land, by 10 percent.[74] Given the centrality of cereals and sugar beet for livestock husbandry and for the food-processing sector, such a dent could send a tremor down the national economy's spine.

Climate modeling, a relatively late addition to the field of scientific inquiry, keeps developing. As earlier methodologies associated with the Special Report on Emission Scenarios (SRES) gave way to methods based on representative concentration pathways (RCP), more authoritative predictions for specific locales kept surfacing. In the Middle East as elsewhere, ever more nuanced forecasts paint an ominous picture. From megacities by the Arabian Gulf growing too hot for human habitation to

Morocco losing a third of its agricultural productivity; from Turkey's agricultural hinterland becoming barren to Egypt's delta choking under a depleted Nile; from Syria's Fertile Crescent peasants plunging into chaos to seasonal shifts that force Yemeni and Saudi pastoralists to sell off herds—outlooks are grim. As worsening climate conditions impede the region's erstwhile struggle against desertification and a looming water crisis threatens agriculture, communities across the Middle East face uncertain destinies.[75]

But climate change does not descend upon the world in blanket fashion. The consequences which it brings to different territories depend also on socioeconomic, political, and cultural contexts. Analyses of adaptation capabilities, mitigation measures, and vulnerability, in other words, must go beyond aggregate regional and even national overviews. In the Middle East as elsewhere, adequate understanding of the social and political contexts of exposure to climate perils and resilience towards them requires a closer look at inequality.

2 BURNING INEQUALITY

Situated on the eastern flank of the French Riviera, Monaco is one of Europe's most prestigious yacht harbors, where many of the rich and famous keep luxury crafts year round ready for the occasional spin in the Mediterranean. Maritime logistics agencies at the Monaco port are at hand to offer owners a variety of services, from keeping license and insurance paperwork intact through hiring crews, and from purchasing fuel, food, and beverages to making sure that furniture, upholstery, cutlery, and linen are in shape. Working for this glitzy crowd is lucrative, an owner of one such agency told television show *Lifestyles of the Rich and Famous*, but it can also involve weird incidents.

She recounted a time when her company prepared a yacht belonging to a Russian oligarch. The owner and his entourage arrived in late afternoon, boarded the super-yacht in time for dinner, and were preparing for a night in the harbor before departure the next morning for their cruise. At dusk, she recalled, she received a phone call from the yacht. The caller was a secretary to the owner's daughter. Her employer, she said, who was on board the yacht with her baby son, had just discovered that the infant's Baby Sense, the small device which monitors his breathing while he is asleep, was left at home. The yacht was large, with many decks and corridors, and the young mother refused to spend a week at sea unless the nanny could constantly monitor the child. Could the logistics agency help?

The woman at the agency, having catered in her time to all sorts of outlandish wishes, reassured the secretary that this could easily be resolved. One of her staff would go to town immediately, buy a new device, and deliver it on board within the hour.

Minutes later, another phone call from the secretary. The mother, she said sheepishly, rejects the notion of a device her son is not accustomed to and insists on having the very one they left at home. Left with no other option, the agency owner dispatched a private jet, which flew from Monaco to Moscow, met a chauffeur who brought the device from the apartment to the runway, and flew back again before the night was out. The device, worth perhaps 50 euros, was on the yacht by dawn.

An hour's flight in a standard eight-seater private jet emits approximately seven tons of CO_2. The return trip to Moscow, which probably involved the plane being airborne for seven hours, emitted some 50 tons of CO_2, an amount equivalent to the yearly emissions of 3 average Americans, 15 Egyptians, or 150 Sudanese.

Climate inequality has three dimensions. First, since different lifestyles generate different quantities of greenhouse gas emissions, those consuming more also carry more responsibility for the advent of climate change in the first place. Second, some are better equipped than others to withstand the perils of the post-normal climate era. Third, those opposing efforts to curb global warming are putting others, including those more vulnerable than they are, in harm's way.

These three components of climate inequality exist everywhere. The Middle East, however, sees them playing out in a particularly striking manner. First, unequal responsibility. The combined population of the 23 countries of the Middle East and North Africa[1] in 2016 was about 550 million, approximately 7 percent of the planet's population. That year countries in the region were responsible for emitting 3.5 billion tons of CO_2, roughly 8 percent of the global amount.[2] Dividing the region's share of world emissions (8 percent) by its share of the global population (7 percent), yields 1.14 as the region's Carbon Inequality Index.[3] An index of 1.0 denotes a territory whose proportion of world emissions are on a par with its global demographic weight. At 1.14, the Middle East's score is only marginally above the global average. Per capita emission figures point in the same direction. In 2016 the average emission for the Middle East and North Africa was 6.4 tons of CO_2 per person per annum, almost the same as the global average of 6.27 tons. Less than half the average per capita emission for the US, it was 20 percent lower than China's, 20 percent higher than the UK's, and many times the average emission in the world's poorest nations.[4]

Average overall emission figures for a region like the Middle East, however, are misleading. Extending from North Africa to Central Asia, the region harbors immense gaps between wealthy states and poor ones. Exposed to similar climate conditions, neighboring countries can nevertheless display huge disparities in consumption levels and, concomitantly, emissions.

In 2013 Kuwaiti citizens had the world's highest average per capita emission, 55 tons CO2e per annum.[5] Qatar, with 37 tons per capita, ranked fourth; Oman was sixth, Bahrain seventh, the UAE ninth, and Libya tenth.[6] Oil-exporting Middle Eastern countries, in short, occupied six of the top ten spots on this dubious global scale. Meanwhile, citizens of poorer countries in the region remain responsible for the smallest quantities of greenhouse gas emissions in the world. Average annual per capita emission in Sudan in 2013 was 0.13 tons, 184th and lowest in the world.[7] Palestine, not featured in the table, had per capita emissions in 2013 of 0.6 tons, a figure that would have earned it rank 180. Yemen, with an average annual per capita emission of 1.4 tons, ranked 158; Morocco (2.33 tons per person) ranked 131; Egypt (3.0 tons) ranked 114;and Jordan, Syria, and Tunisia ranked 104, 109, and 111, respectively . Seven other countries in the region (Israel, Iran, Iraq, Cyprus, Turkey, Lebanon, and Algeria) were ranked between 29 (Israel) and 92 (Algeria), with annual per capita emissions ranging from 11.5 tons per capita (Israel) to 4.9 (Algeria).

So while the 1.14 Carbon Inequality Index for the region is almost on a par with the global average, analysis of individual countries tells a considerably more complex story. Kuwait, with a Carbon Inequality Index of 8.6, is responsible for emissions nearly nine times its proportional weight of the world's population. Qatar is responsible for six times its demographic weight. The UAE, Oman, Libya, and Bahrain emit almost four times their weight. Saudi Arabia, with a Carbon Inequality Index of 2.8, Israel (1.8), Iran (1.5), and Iraq (1.3) also punch above their share of global population. In contrast, the region's 13 poorest countries are found on the opposite side of the scale.[8] Sudan's Carbon Inequality Index (0.05), one of the lowest in the world, is in fact 180 times smaller than Kuwait's.[9]

Another metric for emission inequality is emission intensity, denoting the amount of CO2 emitted per unit of economic activity and expressed as either kilograms of CO2 per dollar GDP or tons of CO2 per million-dollar GDP. Using this yard stick, *The Shift Project Data Portal* recently compared seven global regions: Africa, Europe, Central and South America, North America, Eurasia, Asia/Oceania, and the Middle East.[10] It turns out that between 1981 and 2014, six of the seven saw emission intensities in their respective energy sectors declining. Europe's fell by half, North America's by a third, Africa's by a fifth. The only exception was the Middle East, where the emission intensity of the energy sector surged by a staggering 40 percent.[11]

Here again aggregate regional figures must be examined alongside those of individual countries. While the emission intensity of Kuwait's energy sector went up by 44 percent between 1992 and 2013,[12] that of Jordan dropped by 42 percent.[13] Significantly, this disparity is not unique to the energy sector. CO_2 emissions per dollar GDP in wealthy Middle Eastern countries are roughly double the intensities known in the region's poorer nations.[14] This comparison is even more dramatic when we recall that two-thirds of the region's GDP is generated by only nine countries[15] whose combined populations account for under a third of the region's population.[16]

Emission inequalities exist also within countries.[17] A recent study by French economists Lucas Chancel and Thomas Piketty found that those belonging to the top 10 percent of global emitters were responsible for almost half of planetary emissions.[18] Those in the bottom half of the emissions scale accounted for only 13 percent.[19] A study by Oxfam likewise suggested that average CO_2 emissions per person belonging to the upper global income decile was 11 times the average amount emitted by someone in the poorer half of the world's population.[20] Average per capita emission in the upper global decile, the Oxfam study shows, is 60 times the average emission attributable to an individual belonging to the poorest global decile, a figure congruent with another finding by Chancel and

Piketty. In 1998, they indicate, only a third of global emission inequalities could be explained by within-country inequalities. Twenty years on, internal gaps account for half of global emissions inequality.[21]

Household emissions are linked, of course, to spending power. Residential electricity consumption is influenced by size of the dwelling and how often air conditioners and other appliances are used. My work on CO_2 emissions by households in Israel, for example, indicated that residential electricity consumption of individuals in the upper income decile is 27 times the level of emissions from domestic electricity consumption of individuals in the bottom decile.[22] A similar picture emerges for CO_2 emissions generated by the use of private vehicles[23] and, less drastically, for emissions associated with lifecycle analysis of food consumed.[24] These findings were later corroborated by research in other countries.[25]

All this is highly relevant for the Middle East. A team of prominent economists recently conducted research in the region which used national accounts, household surveys, data on income tax, and a variety of regional and global rich-lists[26] to paint a picture of steep and rigid socioeconomic hierarchies.[27] It turns out that the top 10 percent of earners across the Middle East collect 61 percent of the region's income, almost double the ratio for Western Europe and considerably higher than the figure for the US.[28] Saudi Arabia, Kuwait, the UAE, Qatar, Bahrain, and Oman, home to only 15 percent of the region's population, draw half the region's income. Within those six, individuals belonging to the top decile earn 70 percent of total revenue.[29]

The Middle East has little in the way of a middle class. The 40 percent of earners situated in income deciles two, three, four, and five combined collect among them less than the amount collected by the 10 percent of earners that make the top decile. Not surprisingly, the region has earned itself the dubious title of "the most unequal region in the world."[30]

A report issued recently by Credit Suisse's Research Institute showed that the Saudi economy's worth is $0.7 trillion and that of UAE is $0.6 trillion.[31] The economy of Egypt, whose population is almost four times that of Saudi Arabia and ten times that of the UAE, was valued at $0.4 trillion.[32] Average annual income in the region in 2015 was highest in Qatar with $157,000 per adult, followed by the UAE ($144,400) and Kuwait ($113,400). An average Egyptian that year earned $7,000.

Research into the relationship between socioeconomic inequality and overall emissions in high-income countries suggests that gaps do not only reflect income and expenditure disparities, they also push overall emissions upwards.[33] The share of wealth held by the richest top percent of the population is positively correlated to a country's overall emission figure: the more uneven the distribution of wealth, the higher the overall emission.[34]

This overview of inequalities in terms of CO_2 emissions across the Middle East invites a closer look at the carbon footprints of the region's affluent elites.

The UAE, which hosts some of the richest and most expensive urban centers in the world, provides some striking illustrations. Of the 300 or so private jets purchased throughout the Middle East between 2006 and 2015 to the average tune of $25 million to $75 million apiece, 63 were delivered to the UAE.[35] Aviation industry data suggest that usage of private jets has grown dramatically in the Arabian Gulf area and that a typical aircraft is airborne on average 448 hours a year.[36] Some aircrafts serve their individual owners only. Others are leased to charter companies that farm them out to their own clients, per flight or for a certain number of guaranteed hours over a year.[37] With CO_2 emission for an eight-seater private jet at approximately seven tons per flight-hour, [38] the 63 private aircraft recently delivered to the UAE alone emit almost 200,000 tons of CO_2 every year. Assuming that the number of elite passengers using those jets in the UAE is 1,000, each of them would be responsible for the annual emission of some 200 tons of CO_2 from those trips alone—an amount equivalent to the overall yearly emissions of 60 average Egyptians.

Another source suggests that a UAE businessperson using executive jets flies on average 125 hours per annum[39]—roughly double the average figure for European executives.[40] Assuming that most flights are shared by four persons, the seven tons of CO_2 per hour figure yields an annual emission per person of almost 220 tons of CO_2. Figures for flights onboard full-sized airliners such as the six owned by the Sultan of Oman's household obviously yield much greater personal carbon footprints.[41]

The UAE's largest city, Dubai, had almost 3.4 million inhabitants in 2020,[42] of which 85 percent were non-nationals.[43] According to the World Wildlife Fund's *Living Planet Report,* UAE's per capita ecological footprint for 2008 was the largest in the world, with the United States second and Kuwait third.[44] And while efforts have been made in recent years to steer cities like Dubai and Abu Dhabi towards sustainability, a relentless rise in the standard of living, energy intensive transportation, extravagant construction projects, and norms promoting ostentatious consumption all exact a heavy carbon toll.

Ground transportation is one field responsible for Dubai's staggering amounts of CO_2 emissions. Extended over a large area that is extremely hot most of the year, the city does not invite walking or cycling. With 541 motor vehicles per 1,000 residents and 10,000 taxis carrying almost 150 million passengers per annum, Dubai is easily the most congested city in the Middle East.[45] And while public transport is slowly improving,[46] plans are in place for an additional 500 kilometers of new roads and 120 interchanges.

Another source of monstrous quantities of greenhouse gas emissions is Dubai's series of sea reclamation projects. The famous Jumeirah Palms, constructed in the early 2000s, has 15 leaf-shaped strips connected by a central spine and 10,000 luxury seafront residences, many owned by wealthy expatriates. Two "palms" were added later, as was "The World," 300 man-made islands shaped to resemble the geographical layout of the planet's continents and countries. Reachable from the mainland only by private jet, helicopter, or boat and still in progress, The World had 34 million tons of rock and sand hauled and piled just to create the islets and the 27 kilometer breakwater protecting them from open sea. This alone required staggering amounts of energy and was responsible for immense greenhouse gas emissions.

Ski Dubai is a five-acre indoor facility with a slope boasting a vertical decline of almost 90 meters.[47] According to one source, generating the electricity required to keep the temperature of the snow surface at –16° C and the air inside the complex at –1°C takes 3,500 barrels of oil per day, or 1.3 million barrels a year.[48] Since burning a barrel of oil emits 432 kilograms of CO_2, maintaining Ski Dubai's slopes alone is responsible for the emission of over 500,000 tons of CO_2 per annum,[49] an amount equivalent to the overall emissions of 1.8 million Sudanese.

Another driver of CO_2 emissions in the Middle East is a policy on the part of many governments of heavily subsidizing retail energy prices, thus incentivizing even those with modest incomes to consume lavishly. Whereas the average global price of gasoline in 2019 was $1.13 per liter,[50] filling stations in Kuwait charged 35 US cents. Prices in Bahrain, Qatar, Saudi Arabia, and Oman were around the 50 cents mark while in the UAE gasoline went for 64 cents.

Willing to give up revenue from energy taxation, governments in oil-exporting countries use oil as a tool for indirect wealth redistribution. A tangible perk enjoyed by most on a daily basis, it goes some way to ease the grudge which many hold against outrageous capital accumulation and eye-popping lifestyles on the part of the elite. Not surprisingly, oil-exporting countries in the Middle East lead the world in energy consumed in transportation. In 2014, the average per capita energy uptake in transportation in the Gulf was nearly three times the global figure.[51]

A similar pattern emerges with the price of electricity. Of the 15 countries that have the lowest electricity charges in the world, 8 are Middle Eastern oil-exporting countries.[52] Some have the retail price of electricity as low as 3 US cents per kilowatt/hour.[53] Not surprisingly, Bahrain, Qatar, Kuwait, and the UAE are among the top ten countries in the world in terms of electricity consumed per capita.[54] Other countries in the top ten are Iceland, Norway, Canada, Finland, Sweden, and the US, all of them with substantial territories in the higher latitudes of the temperate climate zone or in the arctic.

The Middle East's polarized reality, whereby elites command legendary riches while others make do with modest incomes, has implications also for overall emissions. The region's major oil exporters in particular seem to accept hyperconsumption, low energy costs for all, and high CO_2 emissions as an acceptable norm. Meanwhile, countries in the region bereft of oil and gas deposits, some of whom count among the poorest in the world, have very low per capita emission levels.

Disconcerting climate forecasts for the Middle East draw attention to the region's vulnerability divide. How will communities respond? Will economic and technological gaps between wealthy communities which hope to endure the post-normal climate era and those who struggle to survive spell salvation for the former and doom for the latter?

Perils associated with climate change were initially regarded as primarily material. More recently, however, researchers and policy makers have been paying more attention to the sociopolitical context of such impacts and the differentiated capacity to withstand them.[55] In 2014, the Intergovernmental Panel on Climate Change's Fifth Assessment Report included a chapter dedicated to links between climate change, livelihoods, and poverty.[56] Identified as "complex, multifaceted, and context-specific,"[57] this nexus has since attracted research into new strategies whereby poverty reduction, climate mitigation, and adaptation measures reinforce one another rather than compete.

A recent G7 working paper entitled "A New Climate for Peace" identified climate-related risks that might destabilize states and their social fabrics.[58] One is, of course, climate change–induced competition for dwindling resources, particularly food and water. Households that fail to cope with such scarcities, the working paper argues, are more likely to resort to illicit economic conduct. Water shortage in particular can spark conflicts for rights to sources, storage facilities, and conduits. Rising sea levels threaten communities occupying marginal coastal territories, many of whom are already chronically stigmatized and socially deprived.[59] Extreme weather events can quickly overwhelm institutions and state agencies, exacerbate poverty in vulnerable locations,[60, 61] and carry detrimental consequences for climate-sensitive sectors including agriculture,[62] urbanism,[63] and health.[64]

To assess the impacts climate change might have on crop yields, researchers frame global flows of agricultural production as taking place in a basic linked system (BLS). In such a system, yields depleted in one area can be balanced by intensified production elsewhere. To model such dynamics, researchers multiply the global acreage of a given crop by its average yield per acre. This allows them to then introduce estimates of how climate change might shrink acreage and how crop yields per acre might be depleted in some territories. The result is an updated estimate of total global crops.

Not surprisingly, studies using this methodology suggest that higher greenhouse gas emissions that will further accelerate global warming will cause considerable dents in global crop yields.[65] A less obvious result, however, is increased inequality. With cereal crops, researchers found, milder climate-change scenarios that will trigger moderate declines in global quantities will nevertheless exacerbate existing inequalities between countries. More affluent industrial countries situated in less vulnerable climate zones will accelerate production, meet a larger share of global demand, and keep aggregate global quantities more or less where they had been. Meanwhile, crops in underdeveloped countries, many of them situated in climatically vulnerable areas, will decline, pushing local food prices upwards and instigating deeper malnutrition, disease, and social unrest. A report issued by the

Intergovernmental Panel on Climate Change in 2019 likewise establishes a disturbing connection between climate change and food availability in many parts of the world.[66]

The Middle East is particularly susceptible to such dynamics. With the exception of Turkey and Iran, all countries in the region rely heavily on imported grain.[67] Those situated in the Arabian Peninsula, but also Libya, Lebanon, and the Palestinian Authority, import some 80 percent of their cereals.[68] Wealthier countries can naturally secure their grain almost regardless of cost. Humbler countries might well be looking at chronic food insecurity going forward.

Adaptation to climate-related water shortage is another field where inequalities play a major role. A comparison between the coping strategies of Saudi Arabia on the one hand and Syria on the other presents a vivid illustration.

With very little rainfall, Saudi Arabia's rain-fed water cycle provides less than 90 cubic meters (m^3) per capita per annum.[69] Well below the global average,[70] this amount cannot sustain rivers or lakes. Moreover, with a considerable proportion of the rain lost in flash floods to the sea, precipitation fails to replenish subterranean aquifers, the all-important water-carrying rock formations that feed springs and wells. This notwithstanding, Saudi Arabia, which many erroneously imagine as an endless expanse of sand and barren rock, developed a thriving agricultural sector since the oil boom of the 1970s. In 2015, farming across the kingdom consumed 14 million cubic meters of water, more than treble the combined amount consumed in the industrial sector (1.2 million m^3) and in private homes (3 million m$^{3)}$.[71] Overall demand for water in the kingdom doubled between 1980 and 2010[72] and has been growing even quicker since.

Saudi Arabia found four main methods to balance its water deficit which could stand it in good stead even in the post-normal climate era. First, the government constructed more than 200 dams since the 1980s which capture flash floods in reservoirs feeding a network of irrigation conduits. Second, the government invested in "fossil water," ancient nonrenewable reservoirs hidden deep in sedimentary rock formations. Drilling is expensive and technically challenging, and sifting the brackish liquid, often mixed with silty particles, requires technologically demanding filtering. Persistent effort in this realm, however, raised the share of fossil water from a third of the national water supply in 1980 to over two-thirds in 2005.[73]

A third tactic in the Saudi water strategy is desalination, used primarily to provide water for domestic use in cities. The capital Riyadh, for one example, gets desalinated water in a pipeline stretching 460 kilometers from the shores of the Arabian Gulf. Finally, Saudi Arabia is a champion of "virtual water," leasing huge productive tracts in Africa and Southeast Asia to secure long-term procurement of imported staple foods.

By the mid-1990s, success in capturing runoff water, desalination, and tapping fossil water reservoirs turned Saudi Arabia into a major exporter of wheat.[74] In less than 20 years, the crown's strategy of ending the country's dependence on imported staples yielded a quadrupling of irrigated acreage. And while accelerated food imports from land leased abroad in recent years reduced the local agricultural sector, fruit, vegetable, meat, and dairy farming are still expanding.[75]

The Saudi success in overcoming chronic water shortage was facilitated by the country's high fluidity since the oil boom. Such strength, alas, was not available to Syria when it became the victim of a major water crisis.

Split between a Mediterranean climate zone in the west and the arid edges of the great Asian steppes to the east, Syria traditionally received precipitation in quantities that adequately sustained a healthy rain-fed agricultural sector. Fed also by the Tigris and Euphrates rivers, the section of the Fertile Crescent which embraces Syria's central plateau, from the Anti-Lebanon ridge in the southwest through the Turkish border in the north to the Iraqi frontier in the east, has been an agricultural hub for centuries. Substantial national investment in this agricultural belt, assisted in the 1980s and 1990s by international aid, modernized the region's agriculture. By the 1990s northeast Syria had become a major source of employment for millions and a mainstay of the national economy.

Under normal circumstances, Syria's natural water cycle, five times the amount of Saudi Arabia's, could support the country's agriculture sector and secure its food supply indefinitely. Abundance, however, created a false sense of security, holding back investments in maintenance of the complex irrigation infrastructure and arresting the development of contingency plans. And so when Syria was hit by a major water crisis, it found itself completely unprepared.

Indications of the potential harm rainfall fluctuations could bring to modern Syria have been around for half a century. Between 1961 and 2009 almost half the winters qualified as droughts.[76] A severe spell of dry winters in the 1960s devastated cereal farming, forcing farmers to eliminate 80 percent of camel stocks and half their sheep.[77] Another series of droughts between 1998 and 2001 forced more than 329,000 people, including 47,000 nomadic households, to downsize herds or liquidate them altogether. Many endured acute food shortages, growing increasingly dependent on emergency relief for their subsistence.[78]

The most devastating crisis, however, famously came in 2007–2011, when a long and bitter drought locked Syria into a calamity of biblical dimensions—a complex humanitarian crisis unlike anything known earlier in the modern era. It began in the winter of 2006–2007, when rainfall quantities plunged 33 percent compared to normal winters. With precipitation in the Fertile Crescent falling by half and some locations sustaining drops of up to 60 percent,[79] the northeast was hit hardest. Sights not seen since the big droughts of the 1960s and the late 1990s were back. Rain-fed wheat and barley lost 47 and 67 percent of their expected annual yields respectively. The national wheat harvest fell from a projected 4.7 million tons to 2.1 million. For the first time since 1992, Syria was forced to import cereals.

The collapse of the agricultural sector in a country whose GDP per capita in 2007 was already ranked 113th in the world snowballed, and Syria fell into an economic abyss.[80] GDP fell from almost $90 billion in 2007 to $55 billion in 2011.[81] Year-on-year inflation rose from 3 percent in 2007 to almost 90 percent in 2013. Unemployment reached 58 percent in 2015; exports fell from $7 billion in 2007 to $2 billion in 2013; and year-on-year GDP plunged from plus 6.5 percent in 2007 to minus 37 percent in 2011.

Syria's downward spiral began in 2007 but was dramatically exacerbated in the years that followed. In three short years the country saw a toxic mix of ethnic schisms, religious tensions, and misguided interventions by an autocratic, inefficient, often corrupt regime turning crisis into chaos. Dams, reservoirs, pumping stations, pipelines, and control posts became scenes of military showdowns between local factions.[82] Instability and insecurity followed, culminating in an unprecedented wave of Syrian refugees seeking shelter in Jordan, Turkey, and western Europe. I shall come back to that in chapter 3.

The UN's World Economic and Social Survey (WESS) suggests that Syria's calamity must not be seen as singular. Rather, it illustrates a vicious circle that could emerge in many countries as physical deterioration triggered by climate change multiplies preexisting social, economic, and political tensions.[83] In such complex situations, disparities in income and in accumulated assets, differentiated access to political influence, gender, age, race, and ethnicity are likely to play decisive roles in the capacity of households and communities to survive. As a rule, when preexisting inequalities are multiplied by new, systemic perils, weaker groups are the first to be endangered.

The Intergovernmental Panel on Climate Change's Fifth Assessment Report of 2014 warns against climate adaptation schemes emulating a problematic syndrome known in the insurance trade.[84] Insurance companies often prefer to rescue those with valuable assets first, neglecting in the process those with fewer possessions even if the latter are more vulnerable. With natural disasters, the report points out, a similar frame of mind can push aid providers to attend to affluent communities first, disregarding their inherent capabilities to deploy their own resources and political influence to help themselves. The less affluent, on the other hand, are often excluded by the very programs designed to help them in the first place.

Data from across countries suggest that poor communities are more exposed to water scarcity than wealthier ones. With lower incomes and less access to financial credit, their ability to diversify is limited, making them doubly vulnerable to the long-term employment and financial consequences of droughts. Gender inequality plays a role too, particularly where agriculture is a primary occupation and women are the main workforce. Whether cultivating their own turf or employed as hired hands, women often find themselves less able to secure alternative sources of income. Children, the elderly, and those with less access to education, health care, and institutional welfare, not least pastoralists, are likewise more susceptible to victimization.

Emergencies often accentuate dependencies of weaker groups on welfare providers at the community, municipal, state, or international level. In Syria as elsewhere, access to support was unevenly distributed. And while communities often displayed remarkable solidarity based on cultural values and social cohesion, structural inequalities often got the better of such local nodes of strength.

The weekend separating the first and second weeks of the 24th annual conference of the United Nations Framework Convention on Climate Change (UNFCCC) in Katowice, Poland, in December 2018 was marred by a disappointing joint statement by the US, Russia, Saudi Arabia, and Kuwait. The four countries, accounting between them to almost 40 percent of global oil production, proclaimed their opposition to a proposed resolution most other countries backed by which the conference's plenary session "welcomes" a recent report prepared by the Intergovernmental Panel on Climate Change.

The report in question was a short but startling summary of the current state of the climate and the steps that need to be deployed to avoid global disaster. The bottom line was unequivocal: despite repeated warnings and regardless of earlier attempts to curb climate change, the situation was deteriorating. Without "rapid, far-reaching and unprecedented changes in all aspects of society," the report stated, the average global temperature in 2100 could be 3 degrees centigrade hotter than its current level. Since this is double the rise most scientists see as acceptable, the report suggests, the situation demands the international community adopts a simple goal: cutting global CO_2 emissions by 45 percent by 2030.

Leading international scientists, government officials, state administrators, and activists were encouraged by the report's bald language. Many hoped that its somber tone might convince world leaders to take real action. Not so the Saudi delegation. Opposed to the tenor of the report and, even more so, to its logical conclusion, the Saudis did all they could to dilute the resolution's language. To achieve this goal, they resorted to a by-law of the UN Climate Convention which stipulates that to pass in the convention's plenary session certain resolutions must enjoy airtight consensus. Joining forces with the Kuwaiti, Russian, and US delegations, the Saudis proposed alternative language whereby the plenary merely "takes note" of the report.[85] The impasse soon derailed the process, eventually causing a sensible proposal comprehensively endorsed by the majority of delegations to be withdrawn.

Saudi Arabia is not only the Middle East's largest economy, largest oil producer, and second largest territory.[86] It is also a major regional power in the field of international diplomacy. First to enter contractual relationships with Western oil explorers almost a century ago, the kingdom has always played a major role in brinkmanship of international energy pacts, financial deals, and investment networks. Other countries in the region entered those arenas later and still look up to the Saudis for leadership and guidance. Within the context of the UN's Framework Convention on Climate Change, the Saudi stance would certainly be regarded as reflecting that of Kuwait, the UAE, Bahrain, Qatar, and Oman, as well as of Libya, Algeria, Iraq, and even, tacitly, Iran.

This goes way back. The Saudi effort in the 1990s to subvert the composition and eventual ratification of the Kyoto Protocol, the UN's earliest and most effective tool to date to combat climate change, has been described as "legendary."[87] Attempts over the years to dilute, distort, or bury resolutions on climate change which went too far for Saudi Arabia's taste included filibustering, evasion, trading votes in other international conventions, and dangling aid to poorer countries. Even the kingdom's 1995 decision to join the UN's Climate Convention in the first place was recently analyzed as driven by obstructionism: it was not until the Saudis realized that the budding convention was gaining real momentum that they decided to join it, with the deliberate intention of hampering its progress as much as possible.[88]

In the 1990s and the early 2000s Saudi Arabia became a vocal opponent of the robust scientific conclusion that climate change is man-made. Later, when this conclusion had been repeatedly and authoritatively solidified with piles of indisputable evidence, the Saudis adopted a different ploy. If a global climate pact was going to phase out fossil fuels, they claimed, then oil-producing countries must be compensated for their loss of future revenue.

One source suggests that the asking price for compensation to the kingdom alone was $100 to 200 billion per annum.[89]

During an earlier conference of the parties of the UN's Framework Convention on Climate Change in 2004, an environmental action group declared Saudi Arabia as winner of the "fossil of the day" award, a shaming device designed to call out efforts on the part of obstructionist delegations to subvert the convention and prevent it from making progress. A source familiar with the matter recounted later that members of the Saudi delegation celebrated the dubious award, treating it as a legitimate source of national pride and expressly hoping they would receive it as often as possible. It was apparently the best and most reliable means for them to communicate their success in undermining climate action to their superiors back home.[90]

In 2009 an episode took place with more serious implications. The weeks leading to the UN's Framework Convention on Climate Change's 15th annual conference in Copenhagen saw a massive disinformation campaign. Orchestrated on a Russian server, the campaign released troves of hacked emails exchanged between leading Western climate scientists. Overinterpreted, distorted, and decontextualized, the emails were falsely presented as proof that climate scientists were insincere and that their scientific work was questionable. Pouncing on the opportunity, Saudi spokesmen reverted to the "inconclusive science" argument. A few weeks later, during the Copenhagen conference, the Saudis played a role in blocking a historic global deal many hoped would be a turning point in the international struggle to curb climate change.

This chapter began by demonstrating how oil-rich countries, responsible for emitting large quantities of CO_2 per capita, carry a heavier historic responsibility for global warming in the first place. It then presented gaps in adaptation capabilities, showing how economic strength and technological prowess sometimes enable wealthy nations to better prepare for climate calamities, a privilege which poorer nations lack. Finally, it highlighted a Middle Eastern activism divide, whereby oil-exporting countries wishing to perpetuate their profits from oil and natural gas have actively subverted global climate action, disregarding consequences this might have for their more vulnerable neighbors.

3 CLIMATE OF INSECURITY

Situated along the Euphrates Valley in the eastern wing of Syria's Fertile Crescent, the provinces of Deir al-Zour and Raqqa saw their combined population grow from 400,000 in 1960 to almost 2,000,000 in 2000.[1] Much of this development came as a result of the government-sponsored Euphrates Valley irrigation project which gave agricultural productivity a major boost. Cereals, livestock, and cotton became the backbones of an expanding agribusiness scene. By 2004 half the population in the two provinces lived off farming, well above the national ratio.[2]

Growth, however, brought new challenges. A government plan to extend farming into the hills above the river valley largely failed. Meanwhile, the swelling population meant family-owned plots had to be divided among more contenders and were becoming ever smaller. Poor maintenance of reservoirs and irrigation conduits caused leakages and waste, pushing farmers to perform illicit drilling and to use wells and river water in unregulated fashion. The regime's failure to manage the system efficiently became increasingly apparent.[3]

The prolonged drought of 2007–2011 brought these problems to a head. Faced with an escalating water crisis, the government imposed water quotas, curtailed agricultural credit, cut subsidies, and raised the price of fuel. A half-baked rural modernization plan, which included a new irrigation scheme, was put together, but it mainly benefited farms owned by wealthy families. Poorer families could not afford the required investment in new equipment, were stuck with their traditional gear, were penalized for it, and saw the economic viability of their farms rapidly diminishing.[4] Overregulation became ubiquitous. In some areas the government forced farmers to only grow cotton, cereals, and sugar beet if they wanted to secure their water quotas. State-owned companies tightened their control on seed supply, credit, pricing, sales, and distribution. As the drought dragged on, crop yields were decimated, causing smallholders to go out of business. Soon Syria was no longer able to feed itself. Oil discoveries in the northeast brought no relief for locals: the few jobs created went primarily to Alawites from Assad regime's heartland in the west.

The downturn of the regional economy triggered widespread bitterness. "Many Euphrates residents," wrote one analyst, "considered themselves an internal colony of Syria well before the civil war."[5] Resentment was initially expressed in peaceful rallies, with participants demanding agricultural reforms, bigger water quotas, and more seeds. The government responded to the demonstrations with harsh clamp downs, quickly radicalizing protesters who soon began demanding a regime change. Violent clashes continued and the death toll rose. The Free Syrian Army (FSA), initially established to protect protesters from police brutality during rallies, morphed into a paramilitary fighting force.[6] By the summer of 2012 it controlled Deir al-Zour's entire rural hinterland and the lion's share of the city itself, pushing government forces to the outskirts.

The ousting of the Syrian army from the northeast, however, did little to ease the economic hardship which went from bad to worse. This opened a window of opportunity for radical Islam. A Jihadi Salafist militia, Jabhat al-Nusra rose to prominence and declared allegiance to the Islamic State in Iraq and Sham (ISIS). Heralding the demise of the regime as proof of infidelity, it began claiming that the imminent arrival of a latter-day Islamic Caliphate was near. By early 2014 Jabhat al-Nusra effectively controlled most of Syria's northeast. It took Assad's regime almost four years and active military intervention on the part of Russia and Iran to regain control of Deir al-Zour and Raqqa in late 2017.

The crisis in Deir al-Zour and al-Raqqa, which forced many to relinquish farming and seek their fortunes elsewhere, had consequences for Syria as a whole. One estimate suggests that of the half million who were displaced by 2017 from Deir al-Zour province alone,[7] 20 percent moved to parts of Syria controlled by the regime; 25

percent sought refuge in Turkey; many others turned to territories held by the Free Syrian Army or the Kurdish militia; and 13 percent ended up as refugees in Europe.[8] In al-Raqqa province as well, hundreds of thousands were displaced.[9] Meanwhile, supporters of ISIS and of Jabhat al-Nusra from other parts of Syria moved to Deir al-Zour and al-Raqqa in large numbers, adding further instability to the demographic upheaval.

Recent research suggests that the Syrian crisis, while obviously linked to the droughts of 2007–2011, may have been exacerbated by a 2011 move by Turkey to retain a bigger share of the Euphrates water in the Ataturk reservoir upstream.[10] This of course reduced the amount of water available downstream, crippling summer crops in the northeastern provinces of Syria. This interpretation puts more weight on the demise of irrigated summer crops than on the collapse of rain-fed winter crops and suggests that Turkey had more to do with Syria spiraling into chaos than hitherto assumed. But given that the Turkish seizure of more Euphrates water came itself as a response to the extended drought, the link connecting Syria's strife to climate remains solid.

The crisis in Deir al-Zour and al-Raqqa provinces had obviously much to do with the particular circumstances of Syria. But the dynamics it exposed are not unique. In the spring of 2007 Israeli troops patrolling the country's southwestern frontier encountered a new phenomenon that took them by surprise. Small groups of African men and women, mostly in their 20s and early 30s, many of them displaying visible signs of physical exhaustion, emerged from the Egyptian side in broad daylight, walked to the Israeli posts, and quietly gave themselves up. In debriefings they all had a consistent tale. Their country of origin was Darfur in South Sudan. They had left weeks, sometimes months earlier, fleeing hunger, war, and persecution, and covered 2,500 kilometers across Sudan and Egypt, some of it on foot, in pursuit of a new life in Israel. It was a journey with biblical reverberations to what they hoped would be for them a land of bounty.

A 2007 UN report on Darfur suggests the crisis there was partly due to a climatic shift. A long dry spell in the 1980s, part of the gradual expansion of North Africa's desert belt southwards towards the subtropics, decimated agriculture in South Sudan, causing hunger and displacement.[11] A nationwide food crisis was exacerbated by ethno-religious tensions. The government in Khartoum framed non-Arab Darfuris as the enemy. Their needs and rights neglected, scarcity in their midst soon escalated into an all-out violent conflict over water, food, and minerals. Mass murders followed. Hundreds of thousands became refugees. Some of those who had been desperate enough to flee and agile enough to travel for weeks in harsh conditions ended up surrendering to Israeli soldiers at the border.

Syria and Sudan, while different in many ways, share important similarities. Both are among the poorest nations in the world. They both depend on local agriculture for nourishment, employment, and national revenue.

Both struggle with desertification, ethnic fissures, and inefficient, often corrupt regimes. And while the dry spells which struck them in recent decades looked unusual at the time, if climate models for the Middle East are anything to go by, misfortunes on this scale could become the new normal in the region.

The nexus of climate change and insecurity seems to play out as a vicious circle with a standard pattern. A long drought triggers an agricultural crisis and tensions over dwindling resources. Local unrest becomes entangled in wider ethnic and political affiliations and alignments. Violent conflict ensues, displacing and victimizing large segments of the population. But what precisely *is* the role of climate shift in such spirals of destruction? Would Syria and Darfur have avoided chaos had they not been hit so badly by consecutive drought years? Will similar dry spells elsewhere inevitably cause comparable calamities? How exactly does the nexus between climate change and insecurity behave?

For analytical clarity, let me distinguish here between two species of security. One, security with a small *s*, emphasizes direct human experience. People everywhere are cognizant of the availability and price of shelter, food, and water and the strengths and weaknesses of the financial and logistic circumstances of their households; those directly dependent on the natural systems monitor them expertly and are able to detect early warning signs of detrimental changes. Within such contexts, notions such as "food insecurity," "diminished productivity," or "damage to essential habitats" have clear experiential meanings. Directly relevant for a sense of self-reliance, these syndromes are unsettling for people's sense of personal security.

Security with a capital *S*, in contrast, is more abstract and less rational. Focused on the collective rather than the personal, its conceptual basis is embodied in concepts such as "national Security," "Security of the realm," or "the Security services." Those in charge of this type of Security often ordain themselves as defenders against external foes, guardians of the social order, and supreme protectors of the institutions that ostensibly ensure well-being and prosperity. Often enjoying wide consensus, Security operations demand immunity from all types of critique and seek to remain permanently shielded from public scrutiny.

One realm in which *s*ecurity sometimes converges with *S*ecurity is mass migration. This is highly relevant for the Middle East, where climate change is likely to expose millions to temporary or permanent water shortages in the not-so-distant future. By 2050, according to one estimate, available water per capita in the region could plunge to under half its current level,[12] with obvious implications for agriculture and, no less important, for perceptions of its future. Anticipating water and food scarcities, people in rural parts across the Middle East will need to strategize. They could have more or less children, downsize plots, sell livestock, remove children from schools to have their labor boost the household's income, lease their land, or, ultimately, sell it and move away.[13]

Migration involves a variety of push and pull factors. It is a complex field, a machine with many moving parts, and identifying the spark that might ignite it in a given context can be difficult. A report published in 2014 by the World Bank, which looked at climate-related migration from rural parts of Algeria, Egypt, Morocco, Syria, and the Republic of Yemen,[14] tried to characterize how climate change and perceptions thereof shape peasants' cognitions, decisions, and actions regarding relocation.

Six percent of respondents across the five countries, the report indicates, listed drought as the primary reason pushing them or members of their households to leave the rural hinterlands and move to urban centers.[15] Six percent cited climate as a secondary cause.[16] But the report suggests that those initial figures, while indicative, fail to cover the entire spectrum of climate change as a push factor. The reason for migration most often cited by respondents was "pursuit of better livelihood." A rather bland assertion, in rural parts it often serves as blanket term for predictions of agricultural decline. In arid zones, the report's authors argue, such pessimism is almost invariably related to expectations of drier, hotter years ahead. Data from focus groups and interviews conducted for the report support this supposition. In all five countries, respondents still residing in the hinterlands as well as those who relocated to urban centers repeatedly referred to impacts of climate change they have already noticed, often citing them as signs of further things to come.

In the Yemen Republic Hudaya region, respondents explained that as water became scarce, quarrels over wells became ubiquitous. Water traditionally used for vegetables and cereals was now diverted to fields of Qat, a lucrative narcotic which demands much larger quantities of water. One respondent recounted how with summer nights now hotter than before, people do not sleep enough, become fatigued, work less, and lose productivity and income.[17] With less water available for food crops and with dwindling crops, many farmers lost confidence and traction. Those who found themselves approaching bankruptcy saw no option other than to sell and move away.

Two-thirds of Morocco's landmass is susceptible to drought. The agricultural sector, accounting for only one-seventh of the country's GDP, supplies employment to almost half the national workforce, including 75 percent of the working poor. Not surprisingly, respondents to the World Bank survey often associated changing climate patterns with rural out-migration. In Tafilelt, a town situated in Morocco's Oasis belt, a quarter of current inhabitants are migrants who have recently relocated from the rural hinterland. Many of them recalled how even during their own lifetimes, the frequency of dry years surged from one in five years to one in two years.[18] Similar testimonies came from the Draa region, once a thriving source of agricultural trade, which now sees farmers departing at an accelerated pace. A forecast many Moroccan respondents subscribed to was that agricultural output will continue to decline, perhaps by 20 percent within two decades, pushing out-migration even further.

Moroccan respondents who contemplated relocation were cognizant of urban housing shortages. They knew of rampant unemployment in the humbler urban quarters. They harbored no illusions about the competition which awaited them against the local poor and were familiar with the cultural barriers that might impede their integration. These notwithstanding, most households still regard migration as their optimal solution and often pool resources to help youngsters make the move.

In times of war, migration often comes as last resort for people desperate to save their families from violence and thus assumes a communal, collective tenor. In Morocco, Yemen, and the other countries featured in the aforementioned World Bank report, migration is a more gradual, individual process. Involving more consideration, choice, and planning, it has everything to do with people seeking to extend their economic opportunities and give their families better lives.

The term *climate refugees* is an extension of *environmental refugees*, first coined in 1985 to denote "people who have been forced to leave their traditional habitat, temporarily or permanently, because of a marked environmental disruption that jeopardized their existence or seriously affected the quality of their life."[19] In both terms, lack of choice seems to be the linchpin: persons are seen as refugees only when relocation is forced upon them by an encroaching military force.

As early as 2001, the Intergovernmental Panel on Climate Change's Third Assessment Report made several references to "environmental refugee" as part of a review of human vulnerability to climate change.[20] Indexing Asia and Africa as continents where risk of involuntary relocation due to climate change is highest, it recognized how rural populations there were already considering out-migration in response to dwindling food supplies, declining health, and economic downturn.

This early insight, however, failed to solve the ambiguity surrounding the meaning of *climate refugee* as a legal term. In 1951, when the UN established the High Commission on Refugees, it adopted a rather narrow definition of refugees. The dual test it used, which still applies in international law, is that for people to be recognized as a refugees they must (a) be displaced as a direct result of military conflict and (b) relocated to a territory across the international borders of their homeland. Subsequent categories, including "economic refugees," "environmental refugees," and "climate refugees" clearly fell outside this scope. Consequently, those who claimed protection under them were generally denied.

This became painfully manifest in 2008, when the UN's High Commission for Refugees published its position on climate change and human displacement.[21] Adhering to the prevailing UN legal framework, the report provided a somewhat incoherent list of climate events that might, in certain circumstances, be recognized as capable of causing forced displacement.[22] Its bottom line, however, brought the debate back to square one: refugees in climate-related contexts will be recognized only if displaced as a result of climate-induced scarcity instigating military conflict and only if they are forcibly displaced across the internationally recognized borders of their homeland.[23]

The report has since had its justified share of criticism[24] following which it was somewhat modified.[25] A decade on, however, with predictions for climate-induced migration ever more robust, legal clarity on climate refugees remains elusive. What we do know is that forced migration is more likely to take place when biospheric deterioration intersects with economic, sociopolitical, and ethnic tensions. Such dynamics are, of course, prevalent across the Middle East.

At least ten Middle Eastern countries harbor ethnic fissures which, while mostly latent, are always on the cusp of suddenly erupting. Modern Iraq is famously divided between Sunis, Shi'aites, and Kurds. Syria is dominated by an ethnic minority, the 'Alawites. Lebanon is officially organized along sectarian lines, with Christians, two Muslim denominations, and a Druze contingent, and maintains rigid quotas of political positions allocated through these strictly denominational parties. Israel, where 1.4 million Palestinian citizens make up a fifth of the entire citizenry, controls the lives of an additional 3 million

Palestinians who live under de facto occupation in the West Bank and the Gaza Strip. Jordan is split between a local population with Bedouin routes and a large contingent of citizens of Palestinian origins. Yemen's tribal history feeds contemporary animosities, including an extended conflict between the Saleh regime and the Houthis, a coalition of tribal segments financed and armed by Iran.[26] Libya, Algeria, and Morocco have sizeable Berber populations, mostly occupying arid territories in the south. Significant portions of Turkey's large Kurdish minority live in areas sensitive to climate change. Egypt is home to over 30 million Sa'idis, a rural group residing in Upper Egypt, often stigmatized by urbanites as country simpletons destined to do demeaning physical labor. It is also home to Nubians, a smaller but more ethnically coherent Nilotic group with a separate language and a distinct identity, and to half a million Bedouins in the western desert and in Sinai. All three groups are less affluent, live in climatically sensitive rural zones, and are aggrieved by chronic marginalization. Lacking access to valuable resources, their response to a future agricultural downturn due to climate change could see pent-up anger quickly radicalized, with a potential to destabilize relations with the state.

The Middle East, as illustrated earlier, also harbors deep socioeconomic inequalities. Oil-exporting countries on the western shores of the Arabian Gulf in particular feature immense disparities between extremely wealthy elites and others who have much less. Indigeneity likewise endows a small minority with rights and privileges that noncitizens are excluded from. And while European, North American, and Australian expatriates, mostly employed in white collar occupations, are relatively safe by virtue of the rights afforded to them by their mother countries, a large underclass of expatriates from Asian countries, many of them employed in menial jobs, are infinitely more exposed. There too an economic crisis driven by a drastic climate shift could ferment unrest.

Many countries in the region are limited in terms of their ability to cope with complex contingencies. The 2017 World Bank index of government effectiveness, which surveyed 193 countries, suggests the average ranking of Middle Eastern states is 114.[27] Merely eight countries score positively (above zero) on this scale, with only Qatar (ranked 21 in the world) and Israel (ranked 23) scoring above 1.0.[28] Four of the five ranked lowest in the world are Middle Eastern states.[29]

The impact this might have became apparent in Northeast Syria and in South Sudan, where lack of trust in state institutions subverted efforts to contain the hardship, turning local tensions into national calamities. Other Middle Eastern countries too have reasons for concern on this account. On the Political Stability Scale, which ranks 195 countries worldwide, the average rank of Middle Eastern countries is 138. Only Oman (ranked 54 in the world), UAE (ranked 60), Qatar (ranked 63), and Kuwait (ranked 105) score above zero. Sudan, Iraq, Libya, Syria, and Yemen, all of whom rank at the bottom of the global table, score below –2.0.[30]

A dangerous combination of climate vulnerability, sociopolitical sensitivity, and inadequate capacity on the part of Middle Eastern governments to properly anticipate adversities and prepare for them does not bode well for the region. If climatic push does come to shove, many in the Middle East might see threats to personal and familial well-being quickly converging with comprehensive risks to Security writ large.

Critical social theorists warn against allowing idioms of Security to create vocabularies of fear which justify repression, a dynamic labeled "securitization."[31] Once securitized, the argument goes, a field can soon become replete with extraordinary government-imposed measures which can then easily become routinized and normalized.[32]

Some have suggested that associating climate change with Security concerns amounts to such securitization.[33] They argue that when institutions such as the CIA, the US military, or the US State Department frame climate change as a strategic threat, they open avenues for the military-industrial complex to demand even more influence and resources than the disproportional amount they already command.

Concerns regarding securitization are warranted. With climate change, however, it can perhaps become a blessing in disguise. Climate change remains a hard sell in many political and public circles, which enables politicians to drag their feet about combatting it. Given this disadvantage, imbuing climate change with a Security angle could inject more urgency into the debate and thus politicize it. The rational and disciplined character of climate modeling—its use of quantified data, consistent procedures, and simple rules of probability—can perhaps avert the pitfalls of the hyped-up, manipulative nature of Securitization that theorists correctly warn against.

Some critics draw particular attention to the risk that Europeans, who have been securitizing incoming migration from the south and east for at least three decades now, will frame the Middle East and North Africa as hot spots of racialized climate refugees who must be kept away. Combined with an apocalyptic view of climate change, such attitudes could further entrench the image of those countries as a perilous, undifferentiated "smooth space," whose passive, apathetic populations threaten the safety of the EU's "striated space."[34]

A relatively recent book on climate change, migration, and security with special reference to Morocco as a "transit state" for climate refugees from deeper Africa partly adopts this critical perspective.[35] Rather than turning against climate-induced migration as a Security threat, it argues, states such as Morocco, which regularly sees migrants from the African Sahel traversing it from south to north, should be encouraged to cooperate with international climate-mitigating efforts. Likewise, the UN's Framework Convention on Climate Change should concentrate on helping vulnerable countries with adaptation measures, thus helping would-be climate refugees from leaving in the first place.

Another suggestion in this context is to strengthen the resilience of vulnerable populations.[36] Communities in climate-sensitive areas, this logic goes, should be encouraged to dig into their own reservoirs of economic, social, and cultural assets, and get assistance from wealthier nations as they endeavor to strengthen their resilience. The danger here, of course, is that "resilience" places would-be climate migrants in a spot long favored by neoliberal ideology whereby they are expected to provide their own salvation. An imagery could thus emerge of "good" dynamics, whereby communities in peril pull themselves out of trouble by their bootstraps, versus "bad" dynamics, where passive communities resist self-reliance or are incapable of it and must hence be shunned by Western biopolitics and kept away indefinitely.[37]

Some view planned relocation of potential climate refugees within their countries of origin as the solution. If large-scale, climate-related displacement in Asia, Africa, and the Pacific is inevitable, this logic goes, why not proactively design preemptive voluntary relocation plans for vulnerable communities?[38] Designed to happen within the mother country, such relocation plans should be treated from the start as permanent, not temporary. A good proportion of their cost should be covered by a global fund financed by wealthy nations.[39] As a first step, advocates suggest, the UN's Framework Convention on Climate Change must compile a protocol for recognizing populations as climate refugees, identify the conduits available for offering them support, and develop the technical specifications relevant for putative relocation efforts.

Interestingly, some countries in the Middle East have embarked on rural resettlement projects even before the advent of climate change. When the High Dam in Aswan was built in the 1960s, Egypt had thousands of Nubians removed before their villages were submerged under Lake Nasser. Similar relocation schemes in the 1980s and 1990s involved smallholders, sharecroppers, and even semi-urban populations from the Nile valley and the Nile Delta who were offered land, small loans, and help with infrastructure if they moved to new settlements in Egypt's periphery. A small wave of rural circulation ensued, with people moving between the East Delta, the West Delta, the Qatara depression, and North Sinai.[40]

Granted, many of those projects fell prey to misconceptions and mismanagement. Incentives were often inadequate and short lived. Recasting social fabrics proved elusive. Facing environmental hazards in the new locations, such as soil erosion and water salinization, many of those enticed, cajoled, or forced to relocate later moved to Egypt's rapidly expanding urban slums.

We could of course judge these relocations schemes and critique the notion of proactively resettling potential climate refugees as neoliberal ploys. Meanwhile, climate change is unrelenting. Future decades will see millions across the Middle East lose their livelihoods, their ways of life, and indeed their lives to global warming. Fortunately, the process will be gradual. So rather than wait for the emergency to actually unfold, planners can still think of how to help the would-be victims help themselves.

Mitigation must obviously remain a top priority, and a range of adaptation tools must be deployed. But territories where agricultural demise appears inevitable, where coasts can no longer be saved, where water is likely to run out, and where communities might already be condemned need relocation plans without delay. The imperfections of earlier resettlement schemes must not serve as deterrents. Rather, lessons gleaned from them should inform current thinking. The debate regarding climate-induced migration and climate refugees is not an intellectual pastime or a theoretical musing. It is about the moral, political, and social imperatives the impending post-normal climate era is already dictating.

4 SOLAR PROSPECTS

The peculiar geology of the Middle East, the role of oil and gas in its recent history, and the troubling climate future which awaits the region imbue its fate in late modernity with poignant irony.

In a distant geological past, the area we now know as the Middle East had decaying plants and animals submerged in seas and lakes. Buried for hundreds of millions of years, some of this organic matter was trapped in underground cavities, morphing into liquefied and gaseous reservoirs of highly concentrated energy. Then in the 1950s, as industrialized nations searched for energy to drive their motors, feed their furnaces, and fire power plants, those ancient products of the region's stratified sedimentation assumed international gravitas. Easily extracted and relatively cheap, oil from the Middle East became a linchpin of the world's economy. Of 135 billion tons of oil ever produced worldwide by 2009,[1] half came from the Middle East;[2] burning this amount generated a third of greenhouse gases human activity ever emitted to the atmosphere.[3]

As data on climate change pile up, the Middle East, already warmer than most regions of the world, emerges as particularly vulnerable. In a peculiar twist of cosmic destiny, the area that contributed so much greenhouse gases to the atmosphere now finds itself on the receiving end of climate's wrath, about to have a bitter taste of its own medicine.

The nexus between modernity, energy, and climate developed in five sequential phases. First came mechanization; second, the fossil fuels required to propel it assumed global economic import; third, material growth occurred on a scale the world had never seen before; fourth, climate change emerged, the unintended consequence of the age of fossil fuels. Finally, with global warming now upon us, humanity faces a fateful choice: stick to fossil fuels and watch the planet's climate spin out of control, or transform electricity generation, transportation, and even industry in a last ditch effort to halt climate chaos.

This five-staged history is associated with a variety of processes, institutions, and connections on global, national, and local scales. Nowhere, however, did the tensions, contradictions, and dilemmas associated with it emerge as vividly as in the oil-rich kingdoms on the western shore of the Arabian Gulf. Saudi Arabia, Kuwait, the United Arab Emirates, Qatar, Bahrain, and Oman have all seen an accelerated version of this chronology.

Mechanization arrived at the Arabian Gulf relatively late but spread across it at breakneck speed. Abu Dhabi, the United Arab Emirate's chief city-state, had its first road paved in 1968 to serve less than a hundred motor vehicles;[4] by 2014 almost 1.5 million cars roamed its hypermodern speedways.[5] The UAE, which in the 1960s had a thousand registered motor vehicles, now has more than 3 million.[6] Per capita electricity consumption in Kuwait grew fivefold between 1973 and 2014,[7] a period in which the figure for the US grew by only 50 percent.[8]

Oil, first produced commercially in Pennsylvania in the late 1850s, became a major global force in the 1950s in a process that catapulted the kingdoms by the Gulf to international prominence. Enticed by American oil companies whose interests dovetailed with those of US foreign policy, local chiefs signed off on lavish exploration and extraction franchises for a small percentage of the profit. Using the royalties from oil to finance lifestyles of luxurious opulence, local potentates also managed to improve the lives of most of their compatriots. An ever-expanding consumer culture ensured global demand for oil grew exponentially, making the deal sweeter by the decade. There was a price to pay in terms of deep dependency on the US, mainly in the realm of security. But lenient labor and environmental laws guaranteed that oil companies and local clients minimized costs and maximized profit. The new political economy of oil, in short, did not merely prop up the oil-exporting fiefdoms on the Gulf financially, it pretty much *defined* them.

In 1960 Saudi Arabia, along with other major oil-producing countries, established the Organization of Petroleum Exporting Countries (OPEC), creating an organ that operates as an international cartel.[9] In 1981 the six kingdoms by the Gulf—Saudi Arabia, Kuwait, the UAE, Qatar, Bahrain, and Oman—established their Gulf Cooperation Council (GCC),[10] a platform which allows them to coordinate local, regional, and global energy policies. Saudi Arabia, Kuwait, the UAE, and between 1960 and 2019, Qatar played central roles in OPEC,[11] not least in the organization's self-ordained mission of maintaining high oil prices through strict production quotas.

The new political economy of oil enabled the kingdoms by the Gulf to enjoy one of the greatest economic booms the world has ever witnessed. The gross domestic product of the United Arab Emirates in 2018 was 138 times its 1973 value;[12] Saudi Arabia's GDP leaped 50-fold from 1973 to 2014.[13] To put this in perspective, the US's GDP grew only 14-fold in the same period;[14] Germany's grew 10-fold.[15]

Then the climate started turning. Poised to severely hit the Middle East at large, global warming will impact the Gulf particularly hard. Even if climate change eventually follows milder forecasts, some of the GCC's hypermodern cities, their pride and joy, could be uninhabitable in 40 years.[16]

Finally, the dilemma surrounding the imminent eclipse of fossil fuels in favor of renewables assumes unique dimensions in the Gulf, a region where economic fortunes have always been connected to the fate of oil.

Interestingly, whereas these five vectors played such a central role in shaping the Gulf's history, they all in fact are driven by forces external to the region. The Gulf's geological history was shaped in an unfathomably distant era; mechanization and the ascent of oil began elsewhere; climate change is fundamentally a global process; and the move towards renewables stems from foreign sensibilities many in the Gulf still see as diametrically opposed to theirs.

Human civilization has a variety of assets that can stand it in good stead in the fateful combat against climate change. Advanced climate models allow precise scaled-down predictions; technologies exist which can reduce emissions; sophisticated policies are available to incentivize and promote the appropriate technologies; there are effective means to modify behavior through learning, culture, mass communication and on social media, and the list goes on.

Decisions taken now on the expansion and deployment of these assets will shape the planet's future. Clearly, while choices on the part of leaders in dominant countries such as the US, China, Japan, Germany, India, the UK, France, Italy, Brazil, and a few others will carry global implications, moves by humbler ones will have less impact. Within this general rule, however, there could be an exception. Saudi Arabia, Kuwait, the United Arab Emirates, Qatar, Bahrain, and Oman, while not among the world's top ten economies, control almost 28 percent of global oil production and 29 percent of all available reserves.[17] This could endow them with a golden opportunity to play a bigger role in this unfolding drama than their sizes warrant.

The Middle East and North Africa were first associated with grand schemes for renewable energy in the 1920s.

One pioneering scheme came from Herman Sörgel, a German architect who envisioned Atlantropa, a gigantic engineering and colonization project in and around the Mediterranean. The centerpiece was going to be a mammoth dam across the Straits of Gibraltar to set the level of the Mediterranean 200 meters below the Atlantic, creating a gradient steep enough for hydroelectric electricity in quantities that would supply Europe's demand for decades. With additional dams across the Dardanelles, between Sicily and Tunisia, and along the Suez Canal, and substantial land reclamation along the shores of the Mediterranean, Sörgel's geoengineering spree was going to transform the region into a global center of energy and food production.[18]

Sörgel's colonial fantasy never materialized. A hundred years on, however, more realistic plans exist to turn parts of the Middle East and North Africa into important hubs of energy production. One recent estimate suggests for example that solar panels covering only a tenth of the Sahara desert can supply the electricity demand of the entire planet.[19] Protagonists of another futuristic project once calculated that even if solar panels were to be constructed only in select locations in North Africa which are particularly suitable, power from them could supply much of Europe's needs by 2050.[20]

Important industrialized countries have already embarked on transition paths which have renewables eclipsing fossil fuels this century.[21] Germany already has 22 percent of its electricity supply generated from renewable sources. More than a third of its renewable capacity, 9 percent of its total supply, comes from solar energy. China's solar capacity as a proportion of its overall supply in 2017 was 9 percent as well,[22] with Italy,[23] Greece, and Spain not far behind.[24] Japan, which had hundreds of golf courses constructed in the 1980s and 1990s only to be deserted later, is now converting many of them to giant solar fields,[25] and the list goes on.

In its early stages, the rapid growth of the renewable energy sector relied heavily on a regime of subsidies and tax incentives. This has largely stopped in recent years. In 2019, firms generating electricity in solar and onshore wind facilities in Europe no longer leaned on EU subsidies. In the UK and Holland even offshore wind farms now make profits without resorting to government support.[26] Offshore turbines in China are also expected to stand on their own financial feet before too long.[27]

In 2004, worldwide electricity produced in renewable energy installations amounted to 227 gigawatts;[28] by 2018 the figure leaped to 1,449. This 640 percent rise in 14 years is visibly reflected in investment volumes. A recent report by Bloomberg indicates that global annual investment in renewables, which in 2004 stood at $32 billion, increased ninefold by 2019.[29] A decade-by-decade comparison suggests that worldwide investment in renewables between 2001 and 2009, $673 billion, more than trebled to $2.18 trillion in the period between 2011 and 2019.

Peak investments in renewables so far came in 2017 and was followed by a 10 percent decline in 2018. The upward trend then resumed in 2019, a year for which producers reported excellent results,[30] and continued into the first quarter of 2020. A Wood Mackenzie Power and Renewables report which came out in February 2020, days before the Coronavirus outbreak, indicated that the first six weeks of the year saw a 15 percent rise in the monetary worth of renewable deals signed compared to the parallel period in 2019.[31] Another report published on the eve of the pandemic predicted that solar power generation in 2020 will be 14 percent higher than the figure for 2019,[32] which would have made 2020 the new peak year for investment in renewables.

The trend towards renewables, particularly solar, does not signal the triumph of environmental ideology. Nor does it reflect some newly found international consensus to halt climate change at all cost. It is premised, instead, on hard financial figures. Lazard, a leading energy consultancy firm specializing in cost-benefit analysis, indicated recently that the unsubsidized cost of generating electricity from solar panels plunged almost 90 percent in a decade, from $359 per megawatt in 2009 to $40 in 2019.[33] The 2019 figure, the Lazard report suggests, is 30 percent below the average unsubsidized cost of electricity from natural gas–fired power plants, which was $56 per megawatt.[34]

The falling costs of renewables, particularly solar, explains another fascinating trend. The period from 2009 to 2019 saw electricity-generating capacity worldwide grow by 2.4 terawatts.[35] Almost two-thirds of this added capacity,

1.4 terawatts, came from renewable energy plants, primarily solar, with wind and hydroelectric in second and third place.[36] In 2020 alone, newly installed generating capacity using solar technologies is expected to reach 142 gigawatts (0.14 terawatts).

Coal- and gas-fired power plants still provide two-thirds of electricity consumed worldwide. The changing landscape of future installation, however, suggests the dominance of fossil fuels in power generation will be short lived. Predictions suggest that by 2050 solar and wind combined will generate more than half of global electricity.[37] Part of this transition will be facilitated by breakthroughs in electricity storage. New technologies are now available that will soon enable smart grids to have mega-batteries feed them at night and during windless spells at costs that will be vastly cheaper than the current ones. No wonder major industrial countries now plan for renewables to become their primary source of energy by midcentury. The prognosis is particularly bright for solar, which could expand its current 2 percent share of global electricity tenfold to 22 percent by 2050.[38]

California provides an early indication of this important trend. In 2019 CAISO, a major grid operator in the Golden State, had renewables, mainly solar, accounting for 20 percent of its supply.[39] During the Coronavirus lockdown of early April 2020, the share of electricity produced from renewable energy in CAISO's mid-morning supply surged to 63 percent.[40] Granted, the lockdown decimated daytime electricity demand from industry, commerce, and offices; April is sunny; and mid-morning is a time of day when solar power generation peaks any day. These notwithstanding, the figure is still staggering and indicates availability, reliability, and an immense potential.[41]

The wrestling match against climate change is, by definition, a race against time. The tipping point where fossil fuels are finally eclipsed must be attained before climate change hits the point of no return, unleashing a variety of feedback mechanisms that could push the biosphere to the abyss. Our collective journey towards renewables is like a train that, having left the station to climb a wooded mountain, is engulfed by a forest fire. The passengers fear for their lives. The engine driver frantically searches the manual for protocols that might enable the train to reach the top unharmed and turn downhill away from the inferno.

The fortunes of the train will be determined by how rapidly the fire spreads but also by the steepness of the slope. Pushing the fable to its limits, the gradient this metaphoric train is struggling against reflects the intensity of the regime of subsidies that props up the universe of fossil fuels. Designed to ensure coal, oil, and gas remain profitable even as the figures stack up against them, this regime is, alas, exceedingly robust. Buttressed by denial and false statements, it dispenses incentives and concessions, grants and tax breaks to some of the wealthiest, most irresponsible corporations the world has ever known.

In the Middle East as elsewhere, influential players with huge interests in fossil fuels have been trying to postpone the demise of oil for decades now. This notwithstanding, leaders in the region are slowly awakening to the potential of renewables to trigger positive change even for them.

Countries in the region whose economic fortunes do not directly hinge on oil and gas exports have long identified renewables as a win-win choice. They realize renewables can emancipate them from oil, a volatile commodity whose cost can shoot up violently and put their national economies at risk. They also know how willingness on their part to embrace renewables can improve their standing with influential countries that lead the global struggle against climate change, including Germany, France, the UK, and the EU at large.

Morocco had 2,760 megawatts[42] of wind and solar capacity installed between 2009 and 2015, increasing the proportion of power generated by renewables from 2 percent to 12 percent.[43] With an installed capacity of 510 megawatts, the Ouarzazate Solar Power Station, known also as Noor (Arabic for light) was for a while the largest power station of its type in the world.[44] Turkey, which has had hydroelectric power supplying a fifth of its entire national demand for decades now, plans a number of solar and wind power plants that will further increase its renewable capacity.[45]

Egypt's New and Renewable Energy Authority (NREA) currently operates some 500 megawatts of wind power plants, with additional 1,340 megawatts under construction. The country also has three private power-generating companies running renewable energy installations which combined provide the grid with almost 2,500 megawatts, and plans exist for an additional 3,200 megawatts to be installed by 2023.[46] If plans materialize, Egypt could have 20 percent of its demand supplied by renewables by 2022, a share that could be doubled a decade later. By 2035, solar could in fact account for 25 percent of Egypt's electricity, wind for 14 percent, and hydroelectric for 2 percent.[47]

Even more significant in the present context, however, is the budding trend in favor of renewables among the region's major oil-producing countries. Iran's renewable capacity more than doubled within three years, rising from 140 megawatts in 2014 to 350 in 2017. With plans afoot for an additional 1,100 megawatts,[48] it is on course to quadruple its 2017 capacity before the end of 2020. Iranian officials have already signaled willingness to export renewable electricity to Iraq, Oman, Afghanistan, Turkey, Armenia, and Tajikistan.[49] Pursuing a similar trajectory, Iraq expects renewables to provide more than 6 gigawatts by 2028 which could account for 10 percent of its aggregate supply.[50]

The six oil-producing kingdoms on the west coast of the Gulf affiliated in the Gulf Cooperation Council (GCC) were, on the whole, slower to embrace renewables than lesser oil producers in the region. In 2014 the six had a combined installed renewable capacity of only 210 megawatts,[51] representing less than 0.3 percent of their overall capacity. By 2018 this amount quadrupled to 867 megawatts[52] but was still shy of 1 percent of their combined capacity.

Of the GCC six, the UAE undoubtedly made the most progress on renewables. As early as 2006 the country's principal emirate, Abu Dhabi, pioneered the Abu Dhabi Future Energy Company Masdar and announced the construction of the future carbon-neutral, zero-waste town Masdar City. Abu Dhabi has since become home to the International Renewable Energy Agency (IRENA) and is the permanent host of the prestigious annual World Future Energy Summit. Consistently promoting solar technologies, the UAE had almost 2 gigawatts worth of renewable capacity installed by 2019[53] and has become a hub for new financial and investment tools specifically tailored for the renewables market.[54]

The UAE's first renewable power plant, with a capacity of 100 megawatts, was commissioned in 2013 in Abu Dhabi. The project provided valuable experience to its stakeholders, including the emirate's Water and Electricity Authority, the aforementioned Masdar, and the Abu Dhabi Regulation and Supervision Bureau. It also served as early confirmation for others, in the region and beyond, that renewables is a practical and economically viable option for large-scale power generation.

Another UAE emirate, Dubai, mobilized its Electricity and Water Authority to initiate the Mohammed bin Rashid Al Maktoum Solar Park. Its first phase, completed in late 2013, had a relatively modest capacity of 13 megawatts. The second phase, which went online in 2017, was initially designed for a capacity of 100 megawatts, but ended up with double the amount. The Spanish firm that won the bid to build and operate it pledged to supply its electricity for $58.8 per megawatt hour, a record low at the time.[55] The contract for phase three, with an additional capacity of 800 megawatts, was won in 2017 by a firm that pledged to sell power to the national grid for only $29.9 per megawatt hour, virtually the same as the cost of electricity from local gas-fired power plants. Construction began in 2017; by May 2018 the project had 200 megawatts installed, with the balance expected to be online by late 2020.[56]

In 2017 the Mohammed bin Rashid Al Maktoum Solar Park contracted a 700 megawatts installation to a Saudi-Chinese consortium.[57] When completed, this plant will boast a solar power tower 260 meters high, the tallest structure of its type in the world, and will employ a state of the art storage facility using molten salt technology. Another UAE emirate, Sharjah, has recently secured financial closure for a waste-to-energy facility, part of a plan to divert three-quarters of its solid waste from landfills to biomass power plants by 2021.[58] In Dubai, 50 megawatts worth of domestic-size solar panels appeared on rooftops, with applications submitted for panels worth 300 megawatts more.[59]

In June 2019, Abu Dhabi inaugurated the world's largest single site solar project in the world. With installed capacity of 1,177 megawatts, Noor Abu Dhabi at Sweihan had $870 million invested in it by the local Power Authority, a Japanese investment bank, and a Chinese solar manufacturer. The electricity goes primarily to the Emirates

Water and Electricity Company for as little as $24 per megawatt hour, another record low for an operating solar project worldwide. At the peak of construction the Indian firm that built the project had more than 3,000 employees on site.[60] In 2018 the UAE's renewable capacity accounted for 3.4 percent of the country's overall production.[61] According to the government's ambitious plan for further installations, this ratio could be doubled by 2025.

Saudi Arabia, the largest, richest, and most influential of the GCC six, is considerably behind the UAE in this realm. Between 2008 and 2018 the kingdom saw energy consumption surge by 60 percent, with demand expected to treble by 2030.[62] The proportion of renewables in this projected growth, however, remains unclear.

When it comes to pledges to expand renewable energy, Saudi Arabia has always had reliability issues. A plan to install 54 gigawatts of renewable capacity by 2030, first announced in the late 2000s, was delayed to 2040 following the global financial downturn of 2008, then shelved for almost an entire decade.[63] Plans for renewables resurfaced only in 2016 as part of the kingdom's Plan for Economic Reform. Vision 2030, a key element of the plan, set a goal of 3.45 gigawatts of installed renewable capacity by 2020. The figure pledged for 2023 was 9.5 gigawatts, which could account for 10 percent of the country's projected electricity supply. The document designated $30–$50 billion to be invested in renewables, mainly in solar and wind.[64]

More pledges came in 2018. In January the Ministry of Energy, Industry, and Mineral resources declared that an investment of $1.5 billion in solar energy would be forthcoming in 2019.[65] Another agency announced plans to have renewables account for 30 percent of the country's installed generating capacity by 2030.[66] In March 2018 plans were revealed for a huge solar power drive, to be executed in conjunction with the Japanese conglomerate Softbank to the tune of $200 billion.[67] Later that year yet another announcement came, this time of a future city on the Gulf of Aqaba that would be powered entirely by renewables.[68] There was talk of 25 gigawatts of renewable capacity in five years, 60 gigawatts in ten, and 200 gigawatts later this century,[69] with investment forecasts ranging from $40 billion to $80 billion.[70]

In April 2019, the research and financial consultancy Wood Mackenzie predicted that Saudi Arabia could be on track to add 6.2 gigawatts of installed wind energy capacity by 2030, a far cry from the 30 percent of all supply announced before. A project involving 850 megawatts scheduled for completion by 2022 was expected to be contracted by the kingdom's Renewable Energy Project Development Office in 2019.[71] The same report however also cautioned that work towards the kingdom's renewables targets for 2030 was already lagging behind schedule.

In reality, by late 2018 Saudi Arabia had only 142 megawatts of renewable capacity installed, primarily in demonstrational solar photovoltaic projects in buildings and in parking lots. This represented 0.18 percent of the country's total installed generating capacity—on a par with Oman and Bahrain, slightly lower than Kuwait and

Qatar.[72] In 2018 Saudi Arabia had an additional 700 megawatts of renewable capacity in various stages of planning, financing, commissioning, and construction.[73] If completed by 2020, these projects could bring renewable capacity to just below 1 percent of the kingdom's total generating capacity.

On the conservation front, in 2018 the Saudi government declared its intention to raise the retail prices of fuel and electricity to a level that would reflect the price of oil on the international market. If fully implemented, this measure could reduce gasoline consumption by 8 percent compared to 2017.[74] Khalid Al Falih, the Minister of Energy, Industry, and Mineral Resources, suggested that the new tariff, combined with forthcoming expansion of renewable capacity, could see daily oil consumption in 2030 down 1.5–2 million barrels compared to current levels.[75]

Finally, whereas Saudi Arabia traditionally designed and executed renewable energy projects exclusively through its somewhat inefficient state-owned investment fund, it has recently liberalized the field by opening it to private sector companies. One consequence was that by 2019 the kingdom was buying power from solar installations for as little as $23.4 per megawatt hour.[76] Analysts are now more confident that should the kingdom want to, it could live up to its declared commitment to renewables.[77]

The GCC six's mediocre record on renewables notwithstanding, recent developments in their respective energy sectors, coupled with ambitious but possibly attainable goals, give some room for optimism. Many across the region now maintain that the transition to renewables makes perfect economic sense even for major oil-exporting countries. This view assumes particular gravitas given the extraordinary potential of the kingdoms by the Gulf for solar power.

Four features common to all the GCC six make them particularly suitable for large-scale solar energy generation plants: abundant sunshine, vast stretches of unproductive land, financial liquidity, and a good track record of integrating innovation into civil infrastructures.

Situated in the heart of the northern global desert belt, the Middle East receives immense amounts of sunlight. With weeks and months of virtually uninterrupted sunny spells in spring and summer, some locations have cloudless skies for over 90 percent of yearly daytime hours.[78] The energy intensity of the sunlight reaching the surface in the Arabian desert at certain times of day is considerably higher than the global average.[79]

Solar power plants use either photovoltaic panels (PV) or concentrated solar power (CSP). In photovoltaic panels, metal plates immersed in a solution are aligned in patterns that have individual plates absorbing different quantities of sunlight. This variance pushes electrons around the plates at different levels of "excitement." This difference in intensity of movement creates a gradient which is then turned into a flow—an electric current.[80]

Concentrated solar power devices, on the other hand, use large lenses or mirrors to focus sunlight into high energy beams. As anyone who has ever used a mirror or a magnifying glass to focus sun rays and burn a hole through paper knows, this type of focused solar energy is immensely potent. If lenses, mirrors, or combinations thereof are large enough, the concentrated beam of energy they form can be intense enough to boil water in high-pressure vessels and create the steam required to turn a large turbine.[81]

The difference between photovoltaic and concentrated solar power is most relevant for the Middle East because of airborne particles. Home to the Arabian and Sahara deserts, two of the largest sandy patches in the world, the region consistently sees huge amounts of airborne sand and dust.[82] Since photovoltaic panels convert direct as well as indirect irradiation, they are largely indifferent to airborne particles. Concentrated solar power installations, on the other hand, rely exclusively on direct sunlight and lose efficiency when a proportion of the sunlight gets deflected by airborne granules. This is why photovoltaic is the preferred technology in most parts of the Middle East. In fact 81 percent of new renewable capacity to be installed around the Gulf by 2030 will use photovoltaic panels. Concentrated solar power installations will account for an additional 10 percent,[83] and wind turbines for the remaining 9 percent.[84]

A second feature which boosts the GCC's potential for solar energy is land. Large-scale photovoltaic fields require vast, easily accessible terrestrial surfaces. The kingdoms by the Gulf hold enormous empty stretches of arid steppe and desert spreads which, bereft of water, are available for more productive uses. In fact, while farmland covers 12 percent of the Earth's landmass,[85] the Middle East has only 4 percent cultivated.[86] And whereas countries in temperate and tropical zones have much of their uncultivated turf covered by forests, the empty quarters of the Middle East are barren.

The third feature is capital. The costs of running coal- or gas-fired power plants are split between construction, a one-off expenditure, and ongoing fuel bills. With solar energy, the main expenditure comes at construction, so demand for capital investment in advance is relatively high. The 12 terawatts of generating capacity from renewables destined to be installed worldwide by 2050 require $13 trillion in new investment.[87] To reach this target, global annual investment in renewables must average $320 billion. It peaked at $298 billion in 2017, went down in 2018,[88] and bounced back in 2019. In early 2020, before the outbreak of the Coronavirus epidemic, it was on track to edge towards the $320 billion mark and possibly beyond.

A closer look at where this capital came from, however, reveals a vast untapped potential by the Arabian Gulf. In 2017, when the combined GDP of the GCC six reached almost $1.5 trillion, some 2 percent of global GDP,[89] their combined investment in renewables, $6.1 billion, represented only 0.2 percent of worldwide investment in renewables.[90] In 2018 combined investment in renewables on the part of the GCC six dropped to $1.24 billion,[91] further lowering their proportional contribution to the global effort.

To reach the global par, the kingdoms by the Gulf must up their game 10-fold compared to 2017 and almost 50-fold compared to 2018. Given the GCC six's financial liquidity, their ability to do this is clearly there. Recent drops in costs could further enhance their motivation to follow this trajectory.[92]

Finally, the GCC six have always displayed a propensity to integrate new technologies into their civil infrastructures. Admittedly, their limited attention to science-based entrepreneurship[93] and low investment in research and training[94] result in unimpressive rankings on global innovation indexes.[95] This notwithstanding, the kingdoms, along with Iran, Turkey, Tunisia, Jordan, and Morocco, have a good track record of technological transformations of their infrastructures and essential civil services, including power grids, water networks, and transportation systems. Artificial intelligence and machine learning, already taught at the graduate level in the UAE[96] and developed professionally in Qatar,[97] could become particularly relevant for managing renewable energy production, storage, distribution, and consumption in the future.

Metrics developed for a future supergrid that will connect all GCC countries and possibly parts of Iraq, Yemen, Jordan, and Egypt and will be supplied primarily by renewables, suggest this grid could save up to 5 million barrels of oil a day.[98] That is equivalent to 1,800 million barrels a year which, if burned in power plants, would produce 800 million tons of CO_2[99]—2 percent of global annual emissions. Should this amount of oil indeed be saved, can we expect that the atmosphere will be spared the greenhouse gases that would have been emitted had the oil been burned? Put differently, if GCC countries convert their local energy sectors to renewables, what will become of the surplus oil and gas which they no longer use?

Three main options present themselves. One is to continue oil and gas extraction at current levels and hope to sell the surplus overseas, perhaps to countries where natural conditions and economic circumstances are less conducive to renewables. This would naturally keep balances of payments on the part of the GCC six more or less intact, but would maintain global CO_2 emissions at current levels with little benefit for climate mitigation.

A second option is to keep extraction at the current level but direct the surplus to industries which use oil as raw materials for fertilizers, food stocks, rubber products, pharmaceuticals, detergents,[100] and perhaps new materials invented and developed in the future. This will, of course, be good for overall emissions, as oil previously burned will now go to noncombustive uses. The potential, however, is limited. Only 7 percent of global oil is currently diverted to such industries,[101] and their ability to take up more oil as raw material is low. This could perhaps change slightly in the future if oil becomes dramatically cheaper, but even then this avenue cannot be assumed to be a game changer.

The third option is the most attractive from a climate point of view: as the GCC six switch their domestic energy sectors to renewables and have their fleets of motorcars go electric, they could decide to reduce oil and gas extraction by the amount now saved. Since most of them already buy electricity from local solar plants at almost the same price they pay for power generated in gas- and oil-fired plants, switching power plants to renewables will not involve major investment. It might even save them some money.

The fate of the excess oil and gas that will emerge when the six kingdoms by the Gulf decide to switch their energy and transportation sectors to solar energy is indeed intriguing. It leads, however, to a bigger question with truly global consequences. Are the GCC six likely to trigger an eclipse of fossil fuel not only in their own backyard but also internationally? Could they deem the global energy transition financially beneficial and politically prudent enough to conclude that rather than resist the eclipse, their own best interests demand embracing it? The concluding chapter raises thoughts on how this might eventually happen.

5 WILL 200 MEN SAVE THE PLANET?

Could their immense potential of solar energy entice the kingdoms by the Arabian Gulf to play an active role in the energy transition now under way across the globe?

For this to happen, three conditions must be met. First, leaders of the GCC six must share a sense of urgency about the need to free their countries from their intense dependency on oil revenues. Second, institutional circumstances conducive to this transition must prevail in global energy arenas as well as at the local level. Third, a coherent plan of action must emerge that will involve as little risk as possible and thus persuade all leaders in the Gulf that failure to act now could be more dangerous than even an incomplete attempt.

The notion that overliance on crude oil sales is unhealthy has been afloat in the region for decades now. Saudi Arabia listed industrial initiatives for oil by-products, primarily feedstocks, in its first five-year development plan in the 1970s[1] and constructed petrochemical plants for fertilizers and other refined oil products the following decade.[2] In 2016 the kingdom's Vision 2030 plan for economic reform,[3] directed personally by Crown Prince Mohammed Bin Salman, declared diversification an urgent national imperative. And while the original blueprint paid scant attention to renewable energy, by 2019 the document was augmented with a pledge of 60 gigawatts worth of renewable capacity to be installed by 2030.[4] Such an amount, which is equivalent to the entire quantity of power generated in the kingdom in 2013,[5] could bring the share of renewable capacity as a proportion of the kingdom's overall power supply in 2030 to a third.

Blueprints for future economic growth issued by the United Arab Emirates followed a similar logic. Vision 2021,[6] UAE Green Growth Strategy, UAE Future Strategy, and UAE Centennial Plan (2071)[7] all emphasize the need to ease dependence on revenues from oil. UAE Energy Strategy 2050 launched in 2017 and since enshrined in law, pledged that by the middle of the century renewables will provide half the country's electricity demand.[8] Kuwait's Vision 2035,[9] originally published in 2010 and revisited in 2017 following a fiscal crisis instigated by falling oil prices, likewise called for diversification. Qatar's National Vision 2030 Development Plan (2008),[10] while primarily preoccupied with modernizing and expanding natural gas resources, declared initiatives for economic activity beyond the energy sector a national priority.

Many initially regarded these vision documents as hollow declarations. More recently however, the notion that economic diversification is imperative has gained more traction as part of the debate surrounding "Peak Oil." Often confused with oil depletion, Peak Oil is a theoretical moment in the future whereby the combined costs of extracting, refining, shipping, and marketing oil could exceed the price consumers are prepared to pay for it. Since costs in different sites diverge, calculations of this tipping point can be specific for an individual field, a country, a region or, in some eventualities, the global industry at large.

Oil and gas reservoirs in the Gulf are found relatively close beneath the surface and are hence easier and cheaper to extract. This gives producers in the region an edge over competitors elsewhere, whose fields often involve negotiating harsh environments, deeper drilling, more complex technologies, and longer distances to refineries, storage facilities, and transportation hubs.

In the past, when demand for oil was always rising, calculations of Peak Oil focused primarily on mounting costs. This gave the oil-rich countries of the Gulf a handsome profit margin, instilling a sense of confidence about their economic future. But things have changed in recent years. With electricity from renewables now more competitive than ever, electric vehicles replacing combustion engine motors, and concentrated solar power entering heavy industry,[11] demand for fossil fuels has grown elastic. This was illustrated clearly in September 2019, when Houthi missiles hit Aramco's oil facilities in Biqayq, crippled half of Saudi Arabia's oil production, and caused a deficit of 5 percent in global oil supply for months. Significantly, the global price of oil stayed put.[12]

Oil's new vulnerability suggests that rather than rising costs, the trigger for Peak Oil could be dwindling demand. This scenario, a nightmare for major oil producers anywhere, conjures up past instances whereby antiquated industries were sidelined by new technologies. Salt, for one example, was used extensively in pre-modern times for meat preservation. In medieval France salt taxes imposed by city-states triggered invasions, displacement, and migrations. Sixteenth-century Poland's economy hinged on the salt trade.[13] Military strategy in the American War of 1812 was partly shaped by a desire to control salt supplies,[14] as were British policies in India at certain junctures. Then in the late 19th century industrially manufactured ice, followed by mass-produced electrical refrigerators in depots, shops and homes, transformed meat storage and changed the food trade irrevocably. Demand for salt was decimated, making the desirable commodity of yesteryear redundant. Tycoons who once had wielded power drifted to obscurity.

Even before Covid-19, the price of oil was too low for countries that heavily depend on selling it. Saudi Arabia, for example, needs oil to fetch $80 per barrel to keep its economy afloat.[15] In late 2019 and early 2020, a price war with Russia brought the price down to $60 per barrel, significantly damaging the kingdom's balance of payments.[16] Granted, with production costs in the region among the lowest in the world, and given the GCC six's robust financial reserves, all of them can stay in business for a while even if oil prices remain low. But even rich economies cannot bleed forever. Unlike commercial companies that can downsize and eventually adjust when markets shrink, states are not so flexible. Dependent as they are on revenues from oil and natural gas, the GCC six cannot sustain themselves forever if profit margins wither.

The perils in store for major oil producers from chronically low energy prices were thrown into relief most vividly in early 2020, with the outbreak of the Coronavirus pandemic known as Covid-19. Even though not the deadliest medical event the world had ever seen, Covid-19 triggered economic, social, and cultural consequences more dramatic than those instigated by any other event in modern history, barring the two World Wars. Never in the history of humankind have billions in all continents been simultaneously threatened by one disease, confined for months to identical routines, and plagued by the same fears and insecurity about the future.

The pandemic's outcomes are pertinent to the nexus of climate change, the Middle East, and Peak Oil on a number of counts—some immediate, others with deeper, longer-term implications

Having disrupted work places, silenced highways, stifled airports, subdued factories, and darkened theaters and stadiums globally, the economic lockdown triggered by Covid-19 dramatically reduced demand for energy,[17] and temporarily dropped CO_2 emissions. Emissions in China in January and February 2020 fell 25 percent compared to previous weeks;[18] daily emissions across Europe fell 58 percent during the 2020 lockdown;[19] New York City, where traffic volumes in 2020 are projected to be 35 percent below average, saw days in March 2020 with half the CO_2 emissions the city normally produces.[20]

As this book goes to print in June 2020, all signs suggest this year will see a sizeable dent in global CO_2 emissions.[21] But banking on this to significantly improve climate predictions for the Middle East or anywhere else in coming decades would be a mistake.[22] As the economy restarts, energy use will increase, pushing emissions up again just as it did after the 2008 financial crisis.[23] A year of mitigated CO_2 emissions is a far cry from reversing a global process that has been rolling out for decades.

A second immediate impact of the lockdown which is even more pertinent to this discussion was a drastic fall in energy prices. Crude oil, long a standard for the industry at large, dropped from $60 a barrel in early January 2020 to $21.44 in early April, before creeping up in May to reach the $40 mark in early June.[24] This exacerbated the problem of diminished revenues for the GCC six described above. With electric engines poised to finally replace combustion motors, long-term demand for oil is vulnerable. In fact when locomotion finally goes electric, power generation will become the only form of energy exchange that really matters. That is why the type of energy that will drive power plants in the future is so crucially important.

If fossil fuels remain as cheap as they were in the immediate aftermath of Covid-19, they might regain the edge over renewables which they had before. Power providers whose fossil fuel bills nose-dive might want to postpone decommissioning their coal-, gas-, and oil-fired plants as much as possible. But plans for new plants, already tilted in favor of renewables, involve long-term considerations that could trump such simple calculations. Dwindling demand and low returns for energy in spring 2020 have already forced many oil, coal, and natural gas providers to reduce production and, in some cases, abandon fields completely.[25] Disrupted sites are complicated and expensive to restart, and the damage to the reliability of supply chains could push these troubled industries further into disarray.

Meanwhile, electricity from renewables, now recognized as financially stable and operationally reliable,[26] is becoming more desirable.[27] Utilities and regulatory boards now tend to lean on them more heavily even if in the short term they are slightly more expensive. No wonder analysts of the global energy market are now suggesting that a post-pandemic recovery could see fossil fuels deteriorating rapidly, perhaps with little hope of ever bouncing back.[28] Like salt a century and half ago, the reign of fossil fuels could soon be over. The recent decline of oil exploration,[29] once a haven of lucrative employment for geologists, seismologists, and engineers, is telling. Those who know the field best, it seems, now act like the proverbial canary birds in toxic coal mines, signaling the end is near.

The oil industry, involving some of the largest companies the world has ever seen, is not likely to take its imminent demise lying down. Using whatever political clout they still have, their captains will mount a bitter struggle to preserve the outrageous regime of subsidies that made them rich for so long. Their effort already faces formidable resistance in Western Europe, where post-pandemic recovery is clearly seen as an opportunity for sustainable development, including quicker transition to renewable energy.[30] On the other hand, as long as Donald Trump is president in the US and Putin and his energy tycoons run Russia, many of the trillions of dollars allocated to recovery will likely prop up fossil fuel providers and high-emitting industries, not least conventional power generation, aviation, and traditional auto manufacturing.

But even if these attempts to work against the tide can buy some time, the age of fossil fuels is almost over. When things do finally go south, the GCC six will need to scramble their way out of a dying business before it chokes them. Their main hope then will be to rapidly divest from oil, cash in on whatever assets they can salvage, and seek new avenues to fortune. The first signs that this reality is dawning in the Gulf came in the spring of 2020, when the Covid-19 related slump in oil prices sent Saudi Arabia on a global investment spree in maritime cruise companies, entertainment agencies, and social media platforms.[31]

The notion that the kingdoms by the Gulf exist on borrowed time is not restricted to their economy alone. It is acutely relevant for their climatic future too. The Gulf, as earlier portions of this book demonstrated, is one of the most climatically vulnerable regions on the planet. Saudi Arabia, where daily summer temperatures already average 32.5°C, could be 6 degrees hotter later this century.[32] Rainfall in its northern parts will grow considerably sparser, with dire consequences for agriculture and local food supply.[33] Summer temperatures in Kuwait could rise by similar proportions, shrinking vegetation cover, accelerating soil erosion, and expanding the amount of airborne particles.[34] The UAE, where 85 percent of the population and most civil installations are situated on shallow beaches along 1,300 kilometers of coastline,[35] is most vulnerable to sea-level rise.Qatar, whose average summer temperatures are the highest in the region, could see them rising to a staggering 40°C by century end.[36] Compounded by extreme humidity, heat stress on the west shore of the Gulf could soon become intolerable a good chunk of the year.

These forecasts assume particular significance given the huge investment recently in new metropolitan centers along the Gulf. Kuwait City, Doha, Dubai, Abu Dhabi, and to a lesser extent Dahran, Manameh, and Muscat were redesigned, expanded, and rebuilt to the tune of hundreds of billions of dollars. Centerpieces of extravagance and luxury, they now personify the prestige and power of the dynasties which sponsored them. Given this symbolic burden, the notion that some of these megacities could be lost to global warming is beyond embarrassing. The very thought could become a liability for the regimes associated with them even before the full brunt of the climate crisis has played out.

For the kingdoms by the Gulf, the advent of climate change and the looming post-oil era are tangible, time-sensitive threats. Pregnant with perilous consequences, each of them separately provides compelling reasons to urgently seek change. Climatically as well as economically, future prosperity in the GCC six might hinge on an early global transformation of power sectors, transportation systems, and even industry to renewable energy. Significantly, global and domestic sets of institutional circumstances are at play which can support such an eclipse.

Accounting for 28 percent of current worldwide oil supply and 29 percent of known reserves, the kingdoms by the Gulf hold a unique position in the political economy of global energy.[37] Saudi Arabia, Kuwait and the UAE, who among them produce the majority of oil originating in the Gulf, have long been pillars of OPEC, a forum wielding much influence in global oil affairs. Admittedly, the tension between Saudi Arabia and Qatar since June 2017, and the exit in 2019 of the latter from

OPEC to join Bahrain, Oman, Russia, and seven other countries in OPEC Plus,[38] somewhat tarnished GCC's cohesion.[39] Also, as *Forbes Magazine* suggested not so long ago, bilateral dynamics between pairs among the GCC six diminish the alliance's unity.[40] These notwithstanding, cooperation among the six in the realms of oil production quotas, sales, and pricing, which builds on decades of common interests, geographical proximity, and cultural homogeneity, is still intact. It was best exemplified in May and June 2020, when coordination among them of extraction quotas buoyed the price of oil from the abyss it fell into during the Covid-19 lockdown. In fact having the GCC six split between OPEC (Saudi Arabia, Kuwait, and the UAE) and OPEC Plus (Qatar, Oman, and Bahrain) probably endows them with even more collective leverage in the global universe of energy.

A second set of institutional circumstances that may facilitate accelerated energy transition in the Gulf is the internal governance structures characterizing the GCC six. Saudi Arabia, an absolute monarchy, has the king as head of state and chief executive ruling by voluntary consultation with a small circle of princes and, occasionally, religious leaders. Kuwait, a constitutional emirate in which the parliament and the judiciary enjoy a degree of nominal autonomy, has a member of the ruling family appointed as prime minister who then selects his cabinet. In the UAE, a federation of seven absolute monarchies, a Supreme Council manned by the seven rulers elects the president and the vice president. The Emir of Qatar is accountable to no one save Sharia (Islamic law), personified in an Advisory Council he himself instates. Bahrain and Oman are likewise administered by cabinets established by their monarchs. None of the GCC six has an electoral system or political parties.

Objectionable in terms of political representation, human rights, personal liberties, and the ability of civil society organizations to make a contribution, autocracies are all about rulers setting policies and implementing them at will. With little in the way of checks and balances, leaders can turn ideas into executive reality without delay. This is reflected in foreign relations too. Reaching consensus among the GCC six probably requires buy-in from no more than 200 men, all of whom are heavily invested in the hierarchies and power structures of their respective dynasties.

Should rulers in the kingdoms by the Gulf conclude that an early global energy transition to renewables is in their own best interests, their plan of action could entail three elements.

First, they could accelerate conversion of their domestic energy and transportation sectors to renewables. This will demand scrapping plans to build new gas- and oil-fired power stations in favour of solar plants and, in select locations, wind turbines, a move already known to make good economic sense. It will need to be augmented by substantial expansion of power storage, so grids can be fed at night, on cloudy days, and during windless spells.

Also, existing fossil fuel plants will need to be decommissioned earlier than scheduled and replaced with renewable facilities. A regulatory landscape geared to encourage rapid conversion of transportation to electric locomotion will have to be instated. High-intensity heat from advanced concentrated solar power (CSP) will need to be gradually introduced to industry. Executed resolutely, the comprehensive transformation of the GCC six's domestic energy realms to renewables could be accomplished in less than a decade.

The second element, to be rolled out simultaneously with the first, is intensive global investment in renewable energy. The GCC six could finance new technologies of power generation and storage; purchase renewable energy generation capacity and storage facilities worldwide; invest in research, development, and production of electric vehicles and smart transportation grids; buy into technologies that apply concentrated solar power in industry; and become champions of financial tools designed specifically to develop renewable energy. It is here that the pressing need to diversify could intersect effectively with the kingdoms' immense economic prowess. Treating renewables at large as a primary growth area and directing their capital reserves toward them could soon land them a solid share of this emerging global market.

The third element, due once the first two are completed, is the most sensitive in terms of planning, timing, and execution. Once they have secured a healthy share in the international renewable energy market, the GCC six could drastically reduce oil and gas production and instigate the final expiration of these industries. Properly coordinated, the dent in global supply their move will have created could nudge the price of oil just high enough to tip the balance permanently in favour of renewable energy, finally establishing it as the cheapest, most convenient and desirable source of energy worldwide.

If they follow this trajectory, GCC leaders could accomplish an extraordinary feat. Rather than cringe at the thought of an inevitable Peak Oil, they will have designed and timed it to suit their own best interests. Making good on their investment in renewables, they will have substituted their erstwhile dominance in the oil trade for an equally robust position in the energy market of the future.

Theoretically, this strategy could be employed by other oil producers too. Any group of countries whose combined stake in oil is large enough could hypothetically begin investing in renewables, then leverage their dominance in both realms to orchestrate a transformation they can capitalize on. Note however that the GCC six share a unique set of climatic, economic, and geopolitical features no other group of countries has. Among major oil producing countries, they are the most climatically vulnerable,[41] the most dependent on oil, and consequently, those most urgently in need of diversification. Compared to other major oil producers, they enjoy the greatest quantities of solar irradiation. Meanwhile, among countries situated like them in the global sunbelt they are by far the wealthiest, hence in the best position to make the investments that could put their immense solar potential into use.

Is the idea that the GCC six might choose to terminate the industry which brought them untold riches naïve and over-optimistic? Perhaps it is. One must never underestimate the tendency on the part of those who benefit from the status quo to glorify it to the point of blocking any type of change. Elites as well as humbler citizens on the western shore of the Arabian Gulf have grown accustomed to an inertial, effortless routine of affluence many of them will be loath to risk. Moreover, with considerable portions of their assets stashed abroad, influential figures in the kingdom by the Gulf must be quite confident that should climatic, financial, or political circumstances turn against them, they will be able to remove themselves, their entourages, and their moveable possessions to cooler, safer locales.[42]

There are, however, two other factors that might help swing the pendulum in the right direction here. First, the international community, desperate for progress in the struggle against climate change, might be willing to remunerate oil-exporting countries for loss of revenue in the event they opt to voluntarily cut production. The notion that wealthy nations could help poorer ones with adaptation and mitigation costs has been part of the climate change debate since the deliberations of the Kyoto Protocol in the 1990s. And while paying oil producers for leaving their assets buried underground has not become a standard feature of the conversation, those who have raised the notion in the past might be inclined to revisit it in future.[43] This ploy in fact is a page right out of the playbook of OPEC, an institution that regularly finds creative ways to reward its needier members for complying with the cartel's decisions to temporarily curtail production levels.

Second, GCC leaders could decide to expedite the end of oil for reasons of honor and prestige. The House of Saud in particular, but other dynasties of the Arabian Gulf as well, regard themselves as flag-bearers of a moderate and modernizing version of benign Islam. Frustrated with what they rightly see as widespread international prejudice against their heritage and culture, they may be partial to some favorable attention. What better way to have such recognition than by becoming part of the solution to a catastrophe some say they have contributed to? Could not their proud association with the culture that invented astronomy, cartography, and algebra convince them to take action that might regain their place of honor in world history?

When all is said and done, a decision on the part of leaders of the GCC six to keep their precious assets buried and embrace renewables requires them to make a mammoth leap of faith. William Durant was born in 1861 to a wealthy Massachusetts family. Grandson to Michigan governor Henry H. Crapo, he entered the carriage-building trade as a young man and by 1900 co-owned the biggest manufacturing company of horse-drawn vehicles in America.[44] When motorcars appeared in the 1890s, Durant was skeptical. He disliked the noise, abhorred the smell, anticipated accidents, and lamented the government's reluctance to properly regulate vehicles and roads. His conviction that the new invention was a menace was so deep in fact that he forbade his daughter to ride in one.[45] By 1903, however, he realized the horseless carriage was here to stay, identified it as an opportunity, and changed direction. Rather than wait for government to impose standards, he decided that he himself would build new cars so clean, secure, and smooth that fears would dissipate completely. He bought a troubled automobile company in Flint, Michigan, called Buick, joined forces with Louis Chevrolet to create General Motors, and the rest is history.[46]

Most carriage makers in Europe and in North America did not survive the motor revolution. Durant, like the Studebaker brothers of South Bend, Indiana,[47] was smart enough to realize that rather than the carriage business, he was in fact part of the transportation trade. No longer willing to let his prior occupation hold him back, he mobilized the infrastructure, reputation, and experience he amassed in the horse-drawn wagons trade to propel himself to motorized glory. Leaders in the GCC now face a similar choice: follow William Durant's example, or end up as most other carriage makers and salt moguls did.

Covid-19 provided other, more positive insights which are relevant for the nexus of climate change, the Middle

East, and the promise of a post-oil era. One has to do with the collective response to a global emergency. For one, the pandemic seems to have restored the place of science in society. Briefings by leaders, parliamentary decisions, executive decrees, media reports, and even discourse on social media during the pandemic had more room for experts and more respect for knowledge than before. Likewise, efforts on the part of medical professionals, researchers, and administrators were acknowledged as essential components of the struggle to safeguard the public's health.[48] As a renowned scholar of environmental law tweeted in late March 2020, "The Covid-19 crisis should teach a lesson that is directly applicable to climate change: heed the warnings of scientists, do everything possible to minimize the hazards they predict and prepare to cope with the impacts that remain."[49]

The initial response to Covid-19 also elevated a sense of community and solidarity at a time when many had already given up on them. Without pomp and ceremony or, for that matter, much deliberation, governments, institutions, and individuals everywhere instinctively prioritized the lives and health of those who are most vulnerable over all else. Economies were downscaled for months, sustaining huge financial loses, in an effort to improve the survival odds of those unfortunate enough to contract the disease. Among them, special attention was given to individuals whose age or preexisting illnesses exposed them to more danger. Paradoxically, leaders who had dedicated their entire careers to destroying the prospect of taxpayer money ever being spent on giving medical attention to those who cannot pay, and who spared no effort to undo financial safety nets for those who found themselves in hardship, now signed off on trillions for these very causes. And while it is too early to determine how much of all resources pledged will find their way effectively to those who need them most, the sentiment reflected in the initial willingness to give was heartening.

The reentry of solidarity into the public domain has to do with another sensibility that is relevant to the climate crisis, namely the pressing need for "muscular, proactive governance" and the responsibility of representative institutions to overpower corporations designed to enrich their owners only.[50] True, some governments demonstrated a propensity during Covid-19 to treat citizens as subjects, keep them isolated, and impose emergency measures that eroded their rights and freedoms. Others at the same time displayed remarkable abilities to enact policy, make good decisions on collective action, shake off defeatism, and help their communities to overcome adversity.

Will emotional, communal, and administrative responses to future climate-related calamities improve as a result of lessons learned from Covid-19? Will accelerated sea-level rise as a result of the rapid breakup of continental ice shelves in Greenland or Antarctica, for example, create the same sense of common global destiny that emerged in the spring of 2020? Will attempts by governments to cope with weakened water cycles and dwindling food supply chains in a warming world mirror the positive features of the struggle against Covid-19?

Most people treat climate change as a distant process, with little or no bearing on their personal well-being. Largely intangible, its encroachment is insidiously slow and indirect, thus difficult to follow and easy to ignore. Covid-19 too has been invisible and omnipresent. Unlike climate change, however, its consequences are direct, personal, discernable, and most importantly, transitory. Pandemics dissipate eventually when a large enough segment of the population becomes immune, when researchers isolate and protect the most vulnerable, or once the cavalry rides in with a vaccine. Climate change, in contrast, has no expiry date. It is here to stay and offers no coherent strategy for exit. Riding the climate storm until salvation comes is not impossible perhaps. It will demand however a mighty global effort that could last for many decades. Perhaps in this respect, the most important lesson Covid-19 has taught us is humility in the face of elements of nature we know little about.

The kingdoms by the Gulf can embrace the opportunity to lead the energy transition, save their skins, and become champions of the global climate struggle in the process. If they procrastinate, let the market run its course against them and join the energy transition under duress, it might prove too late for their economies and, more importantly, too late for the worldwide effort to hold off climate chaos.

Before Coronavirus, the notion that 200 conservative, despotic Arab men could play an active role in stopping climate change was as counterintuitive as was the view, in early 1989, that Gorbachev would undo the Soviet Union or that Mandela would emerge from prison to nullify apartheid and reinvent South Africa. Yet here we are, with the nagging feeling that the current reality in the Gulf, where heat is still just bearable, petrodollars flow with ease, and dynasties look stable, could be short lived. Has the time come for the kingdoms by the Gulf to have a change of heart regarding energy?

ACKNOWLEDGMENTS

This book was conceived when I was on sabbatical leave in New York in 2017–2018. I am grateful to City University of New York's Graduate Center's Advanced Research Collaboration (ARC), graciously headed by Don Robotham and safely steered by Kay Powell, for a fellowship that enabled me to develop the concepts and the data for this book. Special thanks are due to colleagues at the program's Inequality Workshop, whose input greatly helped in shaping chapter 2. Thanks are due also to Boston College's sociologists Andrew Jorgenson and Juliet Schor and their outstanding group of PhD students working on climate change and inequality for a stimulating and most helpful debate in the summer of 2018.

I am indebted to Columbia University's School of International and Public Affairs (SIPA), particularly to my friend Steve Cohen, to Dean Merit Janov, and to Dan Macintyre for the invitation they extended to me to teach a pioneering seminar on Climate Change and the

Middle East in 2017. Cheers to Andrew Angle, Donic Dusseau, Keren Kuperman, Kenny Nguyen, Madeline Ruid, Anna Schimmel, and Saroja Schwager for their attention, thoughts, and creativity during those sessions. Thanks are also due to numerous students in my Sociology of the Atmosphere course at Tel-Aviv University, my academic home, for their enthusiasm and insight.

Neve Shalom's program for Jewish and Arab environmental change leaders, in which I have been proudly involved since 2016, was a great sounding board for ideas. I thank Nava Sonnensein, Mazen Zu'abi, and Noa Barkai for their encouragement and insight. Thanks are extended also to my friend and colleague Khaled Fourani for his thoughtful comments, ideas, and interest, and to Gili Kuperberg for keeping all those files intact.

David Rosen, a fellow anthropologist and friend, kindly invited me to present early versions of some chapters to various forums at Fairleigh Dickinson. I thank him also for conversations in coffee shops in Brooklyn and for his careful readings of various stages of the manuscript. Lia Ettinger, always an inspiration, along with the indefatigable Eilon Schwartz and Alon Tal, generously agreed to read the final manuscript and responded with fascinating feedback and suggestions. Colin Price made useful comments on climate modeling. I thank Eyal Zisser and Amit Mor for fascinating talks and tips on energy and the Middle East, and On Winkler and Avner Wishnitzer, conveners of the Speculative Histories of Climate workshop at Tel-Aviv University in January 2020, for the great discussion they created. Thanks go also to David Lehrer from the Arava Institute for Environmental Studies, Deborah Sandler and Richard Kaplan from Oxford University's Martin School, organizers of the Noam Segal Memorial conference Climate Stress and Regional Risks: The Jordan River Basin in February 2020. These conferences helped greatly in refining some of the book's main points.

Kate Wahl, editor-in-chief and publishing director at Stanford University Press, has been an anchor for this project all along. Her early response and sustained support, proactive commentary, and positive critique were more than I expected or deserve, and have greatly enhanced the scholarship and communicative potential of the book. I am grateful to Caroline McKusick, assistant editor at SUP, and Gigi Mark, production editor, for coordinating the final stages of the project with precision and efficiency; and to Catherine Mallon, whose craft as copy editor enhanced the text immeasurably. I am deeply indebted also to Stanford University Press's anonymous reviewers for their positive critique and their encouragement.

Last but not least, Iros my life partner: for your wisdom, patience, and the loving eyes with which you read me (and so many versions of this work).

—TEL-AVIV, JUNE 2020

NOTES

INTRODUCTION

1. Michael McCormick, Ulf Büntgen, Mark A. Cane, Edward R. Cook, Kyle Harper, Peter John Huybers, and Thomas Litt, "Climate Change during and after the Roman Empire: Reconstructing the Past from Scientific and Historical Evidence," *Journal of Interdisciplinary History* 43, no. 2 (August 2012): 169–220. A fateful drop in rainfall in the sixth and seventh centuries, the article suggests, could explain the Muslim empire's westwards push in the seventh century, as well as the invasion of Europe on the part of Central Asian tribes (""The Barbarians"") and the demise of the late Roman empire.

2. Ibid.

3. Ronny Ellenblum, *The Collapse of the Eastern Mediterranean: Climate Change and the Decline of the East, 950–1072* (Cambridge: Cambridge University Press, 2004).

4. David Kaniewski, Elise Van Campo, and Harvey Weissc, "Drought Is a Recurring Challenge in the Middle East," *Proceedings of the National Academy of Sciences of the United States of America* 109, no. 10 (2012): 3862–3867.

5. Arie Isar and Matanyah Zohar, *Climate Change—Environment and History of the Near East* (Berlin: Springer, 2007), 221.

6. Sam White, *The Climate of Rebellion in the Early Modern Ottoman Empire* (Cambridge: Cambridge University Press, 2013); Geoffrey Parker, *Global Crisis: War, Climate Change and Catastrophe in the Seventeenth Century* (New Haven, CT: Yale University Press, 2013).

7. Dan Rabinowitz, *Here It Comes: How Do We Survive Climate Change?* (in Hebrew) (Tel-Aviv: Hakibbutz Hameukhad, 2009).

8. In 2019 a record 36.8 billion tons of greenhouse gases were emitted into the atmosphere globally, https://www.globalcarbonproject.org/carbonbudget/19/infographics.htm (accessed April 6, 2020). In May 2019 atmospheric CO2 concentration reached the highest ever value recorded, 414 parts per million (0.0414 of a percent of the entire weight of the atmosphere), https://research.noaa.gov/article/ArtMID/587/ArticleID/2461/Carbon-dioxide-levels-hit-record-peak-in-May (accessed April 6, 2020).

9. For one of the earliest indications see C. D Keeling, "The Concentration and Isotopic Abundances of Carbon Dioxide in the Atmosphere," *Tellus*, 12 (1960): 200–203.

10. Studies of impact, adaptation, and vulnerability (acronym IAV) have been promoted by the Intergovernmental Panel on Climate Change (IPCC) since 2009 and involve scientists from a variety of disciplines. See http://sedac.ipcc-data.org/ddc/ar5_scenario_process/parallel_IAV_research.html (accessed October 10, 2018).

11. This book looks at the Middle East at large, including North Africa (the Maghreb) and the Horn of Africa. It uses "“the Middle East”‘ to denote the entire region, making references to particular segments when necessary.

12. The Northern Hemisphere's desert belt stretches from the Atlantic shores of Northern Africa in the west to China in the east.

13. Jason Evans, "21st Century Climate Change in the Middle East," Climatic Change 92 (2009): 417–432.

14. Ibid.

15. For a review of occupations that could cause workers heat stress, see Matthew McKinnon, Elise Buckle, Kamal Gueye et al., *Climate Change and Labour: Impacts of Heat in the Workplace*, April 29, 2016, www.ilo.org. (accessed October 18, 2019).

16. IPCC 4th Assessment Report (2007); for a more recent update on sea level rise see https://www.carbonbrief.org/explainer-how-climate-change-is-accelerating-sea-level-rise (accessed September 20, 2019).

17. Christian Parenti, *Tropic of Chaos: Climate Change and the New Geography of Violence* (New York: Nation Books, 2011).

18. Andreas Malm, "Revolution in a Warming World: Lessons from the Russian to the Syrian Revolutions," *The Bullet*, April 23, 2018.

19. Robin Yassin-Kassab and Leila Al-Shami, *Burning Country: Syrians in Revolution and War* (London: Pluto Press, 2018).

20. J. William Carpenter, "The Biggest Oil Producers in the Middle East," Investopedia, November 8, 2019, https://www.investopedia.com/articles/investing/101515/biggest-oil-producers-middle-east.asp.

21. According to OPEC estimates of oil reserves in 2016, http://www.opec.org/opec_web/en/data_graphs/330.htm (accessed October 18, 2019).

22. "Earth system science," Wikipedia, https://en.wikipedia.org/wiki/Earth_system_science (accessed October 18, 2019).

23. Elizabeth Kolbert, *The Sixth Extinction: An Unnatural History* (New York: Holt, 2014).

24. Dipesh Chakrabarty, "The Climate of History: Four Theses," *Critical Inquiry* 35 (2009): 197–222.

25. Ulrich Beck, "Emancipatory Catastrophism: What Does It Mean to Climate Change and Risk Society?" *Current Sociology* 63(1) 75–88.

26. Bruno Latour, *A Plea for Earthly Sciences*. Keynote lecture, the Annual Meeting of the British Sociological Association, East London, April 2007.

CHAPTER 1

1. Josie Endsor, "Iraq Suspends Rice Farming as Water Crisis Deepens," *The Telegraph*, June 19, 2018, https://www.telegraph.co.uk/news/2018/06/19/iraq-suspends-rice-farming-water-crisis-deepens/ (accessed October 18, 2019).

2. "History of Rice Cultivation," Ricepedia, http://ricepedia.org/culture/history-of-rice-cultivation (accessed October 18, 2019).

3. Endsor, "Iraq Suspends Rice Farming as Water Crisis Deepens."

4. See note 1.

5. Sinead Lehane, "The Iranian Water Crisis," Future Directions International, *Strategic Analysis Papers*, 2014, http://futuredirections.org.au/wp-content/uploads/2014/02/The_Iranian_Water_Crisis1.pdf (accessed November 20, 2019).

6. https://en.wikipedia.org/wiki/2018_Iranian_water_protests (accessed November 20, 2019).

7. Jeremy S. Pal and Elfatih Eltahir, "Future Temperature in Southwest Asia Projected to Exceed a Threshold for Human Adaptability," *Nature Climate Change* 6 (February 2016): 197–200.

8. An area qualifies as having crossed the threshold if it has 35 centigrade wet-bulb temperature lasting for an average of six hours a day or more for an extended period of time.

9. This prognosis was corroborated by J. Lelieveld, Y. Proestos, P. Hadjinicolaou et al., "Strongly Increasing Heat Extremes in the Middle East and North Africa (MENA) in the 21st Century," *Climatic Change* 137, no. 1-2 (2016): 245–260; and by a 2016 report by Max Planck Institute entitled "Climate Exodus Expected in the Middle East and Africa," https://www.mpg.de/10481936/climate-change-middle-east-north-africa (accessed October 18, 2018).

10. Hofit Itzhak Ben-Shalom, "The Combined Effect of Global Warming and the Urban Heat Island and Its Impact on Human Heat Stress Increase in Israeli Cities—Current Trends and Future Predictions" (PhD thesis, The Porter School of Environmental Studies, Tel-Aviv University, 2017). See also Hofit Itzhak-Ben-Shalom, Rana Samuels, Oded Potchter, and Pinhas Alpert, "Recent Trends and Future Predictions until 2060 of Urban Warming in Four Israeli Cities Employing the RegCM Climate Model," *American Journal of Climate Change* 5 (2016): 464–484.

11. If, for example, we could confidently say that global temperatures in the 21st century will increase by 0.2 degrees each decade as it did recently, then we would be certain that, relative to 2000, global temperatures will rise by 2 degrees by 2100.

12. Established in 1988 to provide scientific council to the UN's Framework Convention on Climate Change (UNFCCC), the IPPC (Intergovernmental Panel on Climate Change) is manned by hundreds of scientists seconded by their respective governmental scientific institutions and serves as a clearing house for climate research. It produces an integrative Assessment Report on climate change once every five or six years.

13. One storyline, dubbed A2, assumed '"business as usual"; another (A1) expected intensified industrial production and energy consumption; a third (B1), the "greener" storyline, saw

reduced production and a better use of renewable energy; the fourth (B2) assumed a mixed forecast.

14. In 2000, The IPCC's Special Report on Emissions Scenarios developed the four overarching storylines into a more detailed set of 40 future trajectories. See N. Nakicenovic and R. Swart (eds.), *Special Report on Emissions Scenarios* (SRES) (Cambridge: Cambridge University Press, 2000). SRES storyline A1, which includes 3 energy-use scenarios and 17 emission subscenarios, assumes *technologically efficient, restrained, and equal global growth*: continuous economic growth; new efficient technologies; global population peaking by mid 21st century; and the narrowing of technology, income, and cultural gaps. SRES storyline A2 assumes *business as usual*: slower economic growth, some technological advancement, steady population growth throughout the 21st century, and economic and cultural heterogeneity. Storyline B1, with nine subscenarios, is *environmentally optimistic*. It assumes rapid economic growth but significant reduction in material intensity; cleaner, resource-efficient technologies; global population peaking by mid-century and declining later; technological and cultural global integration; and smaller income gaps. The fourth storyline, B2, which resembles A1, predicts *intermediate, fragmented, and unequal global growth*: slower economic development, technological inequality, continuous population growth throughout the 21st century, and localized solutions to economic, cultural, and environmental challenges that will fail to close global socioeconomic gaps.

15. N. Nakicenovic and R. Swart, *Special Report on Emissions Scenarios* (SRES).

16. Jason Evans, "21st Century Climate Change in the Middle East," *Climatic Change* 92, no. 3 (2009): 417–432. Evans's work is based on the ""business as 'usual" SRES storyline, (A2 in SRES terminology). It assumes that current trajectories of

global demography, production, and development remain at their current levels for the coming decades.

17. Evans, "21st Century Climate Change," 426.

18. Oceans in the Southern Hemisphere, not least the Indian Ocean, produce high-altitude cyclonic systems that transport large-scale volumes of rain-bearing moist air. See also Jason Evans, "Global Warming Impact on the Dominant Precipitation Processes in the Middle East," *Theoretical and Applied Climatology* 99 (2010): 389–402.

19. Ibid.

20. Ibid.

21. Ibid., 430.

22. Filippo Giorgi and Piero Lionello, "Climate Change Projections for the Mediterranean Region," *Global and Planetary Change* 63 (2008): 90–104.

23. The four basic storylines and the ensuing forty subscenarios featured in the SRES methodology were too rigid and difficult to modify in light of real-time climate data. Also, SRES had limited ability to account for future mitigation measures, should those be introduced, and for the impact they could have on global temperatures. Another drawback was that the SRES methodology did not encourage dialogue between researchers in the natural sciences and their counterparts in the social sciences and the humanities, thus further limiting the ability to downscale global models to regional, country, and local levels.

24. R. H. Moss, J. A. Edmonds, K. A. Hibbard, M. R. Manning, S. K. Rose, D. P. van Vuuren, T. R. Carter, S. Emori, M. Kainuma, T. Kram et al., "The Next Generation of Scenarios for Climate Change Research and Assessment," *Nature* 463 (2010): 747–756.

25. The average figure converted to heat at the earth's surface is approximately 1,000 watts per meter squared (W/m^2).

The average figure for the top of the atmosphere is higher (1386 W/m^2).

26. Forcing can either be positive (pushing temperatures up) or negative. It can be internal to the man-made mechanisms driving current climate change, such as increased amounts of greenhouse gases pumped to the atmosphere; or external to this anthropogenic cycle, such as heightened volcanic activity resulting in more greenhouse gases, changes in the ozone hole, or an (unlikely) tipping of the Earth's axis. Radiative forcing denotes the extra amount of energy likely to be trapped as atmospheric heat due to intensified greenhouse effect.

27. The quantity 8.5 suggests that by the year 2100, the extra amount of energy converted at ground level per meter squared and retained in the atmosphere will be 8.5 watts. As a rule of thumb, each extra watt could represent one degree of added global heat. So RCP 8.5 suggests global average temperatures in 2100 8.5 degrees higher than their current level.

28. RCP was later complemented by an additional tool labeled shared socioeconomic pathways (SSP) (See Brian C. O'Neill, Elmar Kriegler, Keywan Riahi, Kristie L. Ebi, Stephane Hallegatte, Timothy R. Carter, Ritu Mathur, and Detlef P. van Vuuren, "A New Scenario Framework for Climate Change Research: The Concept of Shared Socioeconomic Pathways," *Climatic Change* 122 (2014): 387–400. SSP invokes five scenarios, each assuming different future population sizes, economic growth rates, education levels, urbanization, technological advance, and institutional prowess. Two optimistic scenarios (SSP1 and SSP5) assume strong pro-climate action tendencies. Two others (SSP 3 and SSP4) reflect resistance to such action, while SSP2 is in the middle. Using this method, model designers can take a given mitigation target (for example 1.5 degrees of warming by 2100) and calculate the SSP scenario required to achieve it.

29. See for example the International Coordinated Regional Climate Downscaling Experiment (CORDEX), sponsored by the World Climate Research Project (WCRP). In 2012, CORDEX established a body dedicated to regional predictions for the Middle East and North Africa. Driven by needs expressed by regional research affiliates, it facilitated a variety of downscaling projects for many of the region's countries. First simulations emanating from it became available on the Earth System Grid Federation (ESGF) website in 2013. https://esgf.llnl.gov/ (accessed April 7, 2020).

30. IPCC, "Climate Change 2014: Synthesis Report. Contribution of Working Groups I, II and III to the Fifth Assessment Report of the Intergovernmental Panel on Climate Change," R. K. Pachauri and L. A. Meyer (eds.). *IPCC*, Geneva, Switzerland, https://archive.ipcc.ch/pdf/assessment-report/ar5/syr/SYR_AR5_FINAL_full_wcover.pdf (accessed April 15, 2020).

31. IPCC's Sixth Assessment Report due in 2021 will be the first to incorporate SSPs (see note 28) as an integral part of its modeling.

32. "IPCC 2013a: Annex I: Atlas of Global and Regional Climate Projections," in *Climate Change 2013: The Physical Science Basis,* eds. T. F. Stocker et al. (Cambridge: Cambridge University Press, 2013), 1311–1394, https://www.ipcc.ch/pdf/assessment-report/ar5/wg1/WG1AR5_AnnexI_FINAL.pdf (accessed November 20, 2019). The "Atlas" utilizes tools developed by the fifth phase of the Coupled Model Intercomparison Project (CMIP5), a collaborative framework designed to standardize climate predictions and advance research collaboration. By 2013 the project had 43 active models based on this template.

33. Ibid., 1354–1357.

34. Figures presented in the "Atlas" attempt to integrate results from all 43 models known as the CMIP5 ensemble. Each time frame is represented on three maps: one representing

values corresponding to the lower quartile of CMIP5 predictions, one representing the median (50th percentile), and one representing the 75th percentile. The values quoted in the text are for the 75th percentile.

35. Figures pertaining to the 75th percentile of the distribution of the CMIP5 ensemble (ibid.).

36. Mathew Barlow, Benjamin Zaitchik, Shlomit Paz, Emily Black, Jason Evans, and Andrew Hoel, "A Review of Drought in the Middle East and Southwest Asia," *GDIS Droughts Worldwide Special Collection*: 8547–8473 (American Meteorological Society, 2016).

37. Ibid., 8568

38. http://sdwebx.worldbank.org/climateportal/index.cfm?page=country_historical_climate&ThisRegion=Middle%20East&ThisCcode=SAU (accessed October 15, 2018).

39. This rise could reach 30 percent to 80 percent, depending on which model one follows. For interactive projections see http://sdwebx.worldbank.org/climateportal/index.cfm?page=country_future_climate&ThisRegion=Middle%20East&ThisCcode=SAU# (accessed October 15, 2018).

40. Met Office, Climate Observations, Projections and Impacts: Saudi Arabia (UK Department of Energy & Climate Change, 2013),

https://www.metoffice.gov.uk/binaries/content/assets/mohippo/pdf/j/m/saudi_arabia.pdf (accessed August 17, 2019)

41. Ibid.

42. "The United Arab Emirates Second National Communications to the Conference of the Parties of United Nations Framework Convention on Climate Change," *National Communications to the Conference of the Parties of United Nations Framework Convention on Climate Change*, Ministry of Energy, January 2010.

43. UAE, "Climate Change," *Official Portal of the UAE Government*, October 2019, https://government.ae/en/information-and-services/environment-and-energy/climate-change/climate-change, (accessed November 20, 2019).

44. Waleed Hamza, Mohamed Rizk Enan, Huda Al-Hassini, Jan-Berend Stuut, Dirk de-Beer, "Dust Storms over the Arabian Gulf: A Possible Indicator of Climate Changes Consequences," *Aquatic Ecosystem Health & Management* 14, no. 3 (July 1, 2011): 260–268.

45. Thani Al Zeyoudi, "Food Security and Climate Change: The UAE Experience," Arab Forum For Environment and Development, http://afedmag.com/english/ArticlesDetails.aspx?id=116 (accessed November 2, 2018).

46. With wheat crops in central and south Asia projected to shrink by up to 50 percent by 2050, challenges to food security in UAE are destined to get worse. See International Water Management Institute (IWMI), "Facing Climate Change by Securing Water for Food, Livelihoods and Ecosystems," 2017, http://www.iwmi.cgiar.org/wp-content/uploads/2013/02/sp11.pdf (accessed November 2, 2018).

47. Alon Engert and Hanoch Ilsar, *The Ball Is in Our Hands: Global Warming in Israel—Possibilities, Influences and Policy Outlines* (Tel-Aviv: Israeli Union for Environmental Protection, 2007).

48. Assaf Hochman, Paola Mercogliano, Pinhas Alpert, Hadas Saaroni, and Edoardo Bucchignani, "High-Resolution Projection of Climate Change and Extremity over Israel Using COSMO-CLM," *International Journal of Climatology* 38 (2018): 5095–5106.

49. Saaroni Hadas, Baruch Ziv, Judith Lempert, Yael Gazit, and Efrat Morin, "Prolonged Dry Spells in the Levant Region: Climatologic-Synoptic Analysis," *International Journal of Climatology* 35, no. 9 (2015): 2223–2236.

50. Pinhas Alpert, "Climate Predictions for Israel in 2040," Lecture given at the 'Climate Change in Israel' Conference, Jerusalem, September 15, 2009. For the synoptic dynamics that might become the direct causes for reduced precipitation in Israel in the future, see also Assaf Hochman, Tzvika Harpaz, Hadas Saaroni, and Pinhas Alpert, "Synoptic Classification in 21st Century CMIP5 Predictions over the Eastern Mediterranean with Focus on Cyclones," *International Journal of Climatology* 38, no. 3 (2018): 1476–1483.

51. Cook, B., Anchukaitis, J. K., Touchan, R., Meko, D. M., and Cook, E. R., "Spatiotemporal Drought Variability in the Mediterranean over the Last 900 Years," *Journal of Geophysical Research* 121, no. 5 (2016): 2060–2074.

52. P. Gleick, "Water, Drought, Climate Change and Conflict in Syria," *Weather, Climate, and Society* 6 (2014): 331–340.

53. Tom Friedman, "WikiLeaks, Drought and Syria," *New York Times,* January 21, 2014.

54. Gleick, "Water, Drought, Climate Change and Conflict in Syria," 334.

55. http://sdwebx.worldbank.org/climateportal/index.cfm?page=country_future_climate&ThisRegion=Middle%20East&ThisCcode=SAU (accessed December 2, 2018).

56. Martin Dubrovský, Michael Hayes, Pierpaolo Duce, Miroslav Trnka, Mark Svoboda, and Pierpaolo Zara, "Multi-GCM Projections of Future Drought and Climate Variability Indicators for the Mediterranean Region," *Regional Environmental Change* 14, no. 5 (2014): 1907–1919.

57. Samireh Saymohammadi, Kiumars Zarafshani, Mohsen Tavakoli, Hossien Mahdizadeh, and Farzad Amiri, "Prediction of Climate Change Induced Temperature and Precipitation: The Case of Iran," *Sustainability* 9, no. 1 (January 1, 2017): 146.

58. The project used SRES methodology's A2B scenario—a subset of the business as usual storyline.

59. http://sdwebx.worldbank.org/climateportal/index.cfm?page=country_future_climate&ThisRegion=Middle%20East&ThisCcode=IRN (accessed December 2, 2018).

60. https://climateknowledgeportal.worldbank.org/country/turkey (accessed July 6, 2020)

61. Ibid.

62. Ibid.

63. H. Dudu and E. Çakmak, "Climate Change and Agriculture: An Integrated Approach to Evaluate Economy-Wide Effects for Turkey," *Climate and Development* 10, no. 3 (2018): 275–288.

64. For accounts of the geopolitical consequences of riparian relations on the Nile see Ashok Swain, "Ethiopia, the Sudan, and Egypt: The Nile River Dispute," *Journal of Modern African Studies* 35, no. 4 (1997): 675–694; and Haggai Erlich, *The Cross and the River: Ethiopia, Egypt, and the Nile* (Boulder, CO: Lynne Rienner, 2002). For an account of the impact of 'Ethiopia's 2013 declaration of its Millennium Dam Project on tensions with Egypt see F. H. Lawson, "Egypt, Ethiopia, and the Nile River: The Continuing Dispute," *Mediterranean Quarterly* 27, no. 1 (2016): 97–121.

65. Mohammad Hereher, "Vulnerability of the Nile Delta to Sea Level Rise: An Assessment Using Remote Sensing," *Geomatics, Natural Hazards and Risk* 4 (2010): 315–321.

66. Ibid.

67. The delta's western section is vulnerable even to the estimated 1 meter rise predicted by most models for the late 21st century.

68. "Potential Impact of Sea Level Rise: Nile Delta," UN Environment Programme, Grid Arendal, https://www.grida.no/resources/6479 (Accessed December 13, 2018).

69. Ibid.

70. Jane Arraf, "In Egypt, A Rising Sea—And Growing Worries about Climate Change"s Effects," 2017, NPR, https://www.npr.org/sections/parallels/2017/08/13/542645647/in-egypt-a-rising-sea-and-growing-worries-about-climate-changes-effects (accessed April 7, 2020).

71. Ibid.

72. Jane Sušnika, Lydia S. Vamvakeridou-Lyroudia, Niklas Baumert, Julia Kloos, Fabrice G. Renaud, Isabelle La Jeunesse, Badr Mabrouk, Dragan A.Savića, Zoran Kapelan, Ralf Ludwig, Georg Fischer, Robert Rosonf, and Christos Zografos, "Interdisciplinary Assessment of Sea-Level Rise and Climate Change Impacts on the Lower Nile Delta, Egypt," *Science of the Total Environment* 503-504 (January 15, 2015): 279–288.

73. Ibid.

74. Ismail Ouraich, Hasan Dudu, Wallace E. Tyner, and Erol H. Cakmak, "Agriculture, Trade, and Climate Change Adaptation: A Global CGE Analysis for Morocco and Turkey," *Journal of North African Studies* 24, no. 6 (2019): 961–91.

75. Jeannie Sowers, "The High Stakes of Climate Adaptation in the Middle East and North Africa," *Current History* (December 2017): 348–354.

CHAPTER 2

1. Including Sudan, South Sudan, and Palestine, excluding Arithrea, Ethiopia, and Somalia.

2. Greenhouse gas emissions data source: https://en.wikipedia.org/wiki/List_of_countries_by_greenhouse_gas_emissions (accessed January 15, 2019). Population data source: https://en.wikipedia.org/wiki/List_of_countries_by_past_population_(United_Nations,_estimates) (accessed January 21, 2019).

3. For a definition of the Carbon Inequality Index (CII) see Dan Rabinowitz, "Climate Injustice: CO2 from Domestic

Electricity Consumption and Private Car Use by Income Decile," *Environmental Justice* 5, no. 1 (2012): 38–46. Briefly, the CII is calculated in two stages. First, the annual quantity of greenhouse gas (GHG) emitted from a given territory (millions of tons of CO_2 equivalent) is divided by global GHG emissions that year. For example: Turkey in 2016 emitted 408.5 million tons CO_2e—0.9% of global emissions. Next this figure is divided by the territory's population as a proportion of global population. In this example, Turkey's 80.7 million inhabitants represent 1.07% of the global population. Divide Turkey's share in global emissions (0.9%) by its share of the global population (1.07%), and you get its CII (0.84)). A territory whose share of global emissions is equal to its share of the world's population will have a CII of 1.0. Countries whose share of global emissions exceed their weight in the global population will have a CII value higher than 1.0. Those whose share of global emissions is smaller than their share of global population will have a CII value below 1.0.

4. The precise figures for 2016 were: China: 7.36 tons per capita per annum (t/pc/pa); France: 5.16 t/pc/pa; UK: 5.93 t/pc/pa; US: 16.44; Canada: 15.52; South Sudan: 0.13; and Chad: 0.05 t/pc/pa, https://ourworldindata.org/co2-and-other-greenhouse-gas-emissions (accessed January 20, 2019).

5. CO_2e, denoting CO_2 equivalent, indicates that GHG emissions, while dominated by CO_2, also include other gases with varying levels of greenhouse potential. CO_2e contains a measure of the impact of other GH gases emitted calculated as if it came from CO_2, thus presenting a unified, easily comparable figure.

6. https://en.wikipedia.org/wiki/List_of_countries_by_greenhouse_gas_emissions_per_capita (accessed January 25, 2019). The next MENA country in this table is Saudi Arabia, whose per capita emission of 19.2 tons per annum earns it 18th place.

7. Figures for Sudan (2013) were derived from https://tradingeconomics.com/sudan/co2-emissions-metric-tons-per-capita-wb-data.html (accessed January 20, 2019).

8. Middle Eastern countries at or below CII 1.0 are: Cyprus (CII=1.0); Turkey (CII=0.8); Algeria and Lebanon (CII=0.7); Syria (CII=0.6); Jordan, Tunisia, and Egypt (CII=0.5); Morocco (CII=0.4); and Yemen (CII=0.2), https://en.wikipedia.org/wiki/List_of_countries_by_greenhouse_gas_emissions_per_capita (accessed January 15, 2019). No figures were available in 2013 for Palestine and Sudan, who can be safely assumed to be at the lowest end of the CII.

9. Sudan, which emits 12.4 million tons of CO_2 per annum (0.03% of global emissions), has a population of 40 million (0.54% of the world's population)—ICC=0.055.

10. See http://www.tsp-data-portal.org/Carbon-Intensity-of-GDP#tspQvChart (accessed January 25, 2019).

11. From 130 kg CO_2 per $1,000 GDP in 1981 to 180 kg in 2014. Ibid.

12. From 90 kg per $1,000 GDP in 1992 to 130 kg in 2013. See http://www.tsp-data-portal.org/Carbon-Intensity-of-GDP#tspQvChart (accessed January 25, 2019).

13. From 206 kg per $1000 in 1992 to 120 in 2013. See http://www.tsp-data-portal.org/Carbon-Intensity-of-GDP#tspQvChart (accessed January 25, 2019).

14. Oil-producing countries in the Middle East had CO_2 intensities per GDP (for their whole economies) of approximately 400 kg CO_2 per $1000 GDP (2011). Poorer countries in the region typically had approximately 200 kg CO_2 per $1000 GDP. See https://landportal.org/book/indicator/wb-enatmco2eppgdkd (accessed January 26, 2019).

15. H. Plecher, "Gross Domestic Product of the MENA Countries in 2018," Statista, April 20, 2020, https://www.statista.com/statistics/804761/gdp-of-the-mena-countries/ (accessed June 6, 2020).

16. In 2018, the Middle East and North Africa had a combined population of 587 million. The combined populations of the Middle East's nine oil-exporting countries (Iran, Iraq, Saudi Arabia, UAE, Libya, Oman, Kuwait, Qatar, and Bahrain) was 185 million. Figures derived from: http://www.worldometers.info/world-population/population-by-country/ (accessed January 21, 2019).

17. A. Druckman and T. Jackson, "Household Energy Consumption in the UK: A Highly Geographically and Socioeconomically Disaggregated Model," *Energy Policy* 36, no. 8 (August 2008): 3177–3192; Kyle W. Knight and Juliet B. Schor, "Economic Growth and Climate Change: A Cross-National Analysis of Territorial and Consumption-Based Carbon Emissions in High-Income Countries," *Sustainability*, 6, no. 6 (June 1, 2014): 3722–3731; Siqi Zheng, Rui Wang, Edward L. Glaeser, and Matthew E. Kahn, "The Greenness of China: Household Carbon Dioxide Emissions and Urban Development," *Journal of Economic Geography* 11, no. 5 (2011): 761–792; Sai Liang, Shen Qu, Zeqi Zhu, Dabo Guan, and Ming Xu, "Income-Based Greenhouse Gas Emissions of Nations," *Environmental Science & Technology* 51, no. 13 (January 3, 2017): 346–355; Sai Liang, Hongxia Wang, Shen Qu, Tiantian Feng, Dabo Guan, Hong Fang, and Ming Xu, "Socioeconomic Drivers of Greenhouse Gas Emissions in the United States," *Environmental Science & Technology* 50, no. 14 (July 19, 2016): 7535–7545.

18. Lucas Chancel and Thomas Piketty, "Carbon and Inequality: From Kyoto to Paris," Paris School of *Economics,* 2015, http://piketty.pse.ens.fr/files/ChancelPiketty2015.pdf (accessed January 21, 2019).

19. Ibid.

20. Oxfam, "Extreme Carbon Inequality", Oxfam.org, December 2,' 2015, https://www-cdn.oxfam.org/s3fs-public/file_attachments/mb-extreme-carbon-inequality-021215-en.pdf.

21. Chancel and Piketty, "Carbon and Inequality: From Kyoto to Paris."

22. See note 18. Rabinowitz, "Climate Injustice," 38–46.

23. Ibid.

24. Guy Milman and Dan Rabinowitz, "Food Consumption and Environmental Justice: Greenhouse Gas Emissions from Food Consumption in Israel by Socio-Economic Standing," *Israeli Sociology* 18, no. 2 (2017) (in Hebrew).

25. In Indonesia, for example, individuals in households belonging to the upper quintile (the richest 20 percent of households) emit almost eight times CO_2 compared to individuals in households at the fifth (lowest) quintile. Reinforcing the positive association between emission disparities and inequality in purchase power, the Indonesian data suggest that gaps tend to grow along the income scale. While at the lower end emission inequalities decline as income rises, gaps tend to increase as we near the top. Importantly, in Indonesia as elsewhere, the richest 20 percent of households display the highest internal variation, with emissions at the top percentile exponentially higher than anywhere else in the quintile.

26. F. Alvaredo, L. Assouad, and T. Piketty, '"Measuring Inequality in the Middle East, 1990–2016: The World's Most Unequal Region?,"' World Inequality Database: World, Working Paper No. 2017/15, 2017; F. Alvaredo, A. B. Atkinson, L. Chancel, T. Piketty, E. Saez, and G. Zucman, '"Distributional National Accounts (DINA) Guidelines: Concepts and Methods Used in the World Wealth and Income Database," World Inequality Database, Working Paper No. 2016/1, 2016.

27. This effort was important because reliable statistics on wealth distribution and socioeconomic inequality in MENA are difficult to come by. For example, the World Bank global GINI map (https://data.worldbank.org/indicator/SI.POV.GINI?

locations=FR-BH-AF-DZ-AU&view=map [accessed January 12, 2019]) has most Middle Eastern countries (barring Egypt and Turkey) marked as "no data."

28. In Europe, the top decile collects 36 percent of overall income; in the US the figure is 47 percent (ibid.).

29. Ibid. See also L. Assouad, L. Chancel, and M. Morgan, "Extreme Inequality: Evidence from Brazil, India, the Middle East and South Africa," American Economic Association P&P, 2018; World Bank, "Inequality, Uprisings, and Conflict in the Arab World," World Bank and Middle East and North Africa Region; MENA Economic Monitor, 2015.

30. Lydia Asoud Lydia, "Is the Middle East the 'World's Most Unequal Region?" Economic Research Forum, 2018, https://theforum.erf.org.eg/2018/03/27/middle-east-worlds-unequal-region/ (accessed January 27, 2019).

31. All dollar figures denote US dollars; *Saudi Gazette*, "Number of Millionaires in Saudi Arabia and UAE 'to soar in 5 years,'" Al Arabiya News, October 20, 2015, http://english.alarabiya.net/en/business/economy/2015/10/20/Number-of-millionaires-in-Saudi-Arabia-and-UAE-to-soar-in-5-years-(accessed January 28, 2019).

32. Ibid.

33. Kyle W. Knight, Juliet B. Schor, and Andrew K. Jorgenson, "Wealth Inequality and Carbon Emissions in High-income Countries," *Social Currents* 4, no. 5 (2017): 403–412.

34. This trend, which emerged in higher income countries since the 1990s, is absent from low- and middle-income countries (ibid.).

35. Arabian Business Industries, "293 Private Jets Worth $14.6bn Delivered to Middle East Since 2006," Arabianbusiness.com, December 17, 2016, https://www.arabianbusiness.com/293-private-jets-worth-14–6bn-delivered-middle-east-since-2006-655779.html (accessed January 21, 2019).

36. Ed Clowes, "New Blood and Lower Costs Re-energise Exclusive World of Private Jets," Gulf News, November 18, 2017, https://gulfnews.com/business/aviation/new-blood-and-lower-costs-re-energise-exclusive-world-of-private-jets-1.2126629 (accessed January 21, 2019).

37. Private Jet Card, "XOJET and JetSuite Have the Busiest Fleets of Any Private Jet Membership Programs," Private Jetcard Comparisons, August 2016, https://privatejetcardcomparisons.com/2017/07/11/how-many-hours-a-year-do-private-jets-fly/ (accessed February 6, 2019).

38. This figure is for an eight-seater Challenger, which burns 266 gallons jet fuel per hour. See https://www.sherpareport.com/aircraft/fuel-burn-private-aircraft.html (accessed February 7, 2019).

39. Vicky Karantzavelou, "UAE Businessman's Average Use of Private Jets Is 100 to 150 Hours per Year, Against 50 to 100 Hours in Europe," *Travel Daily News*, March 31, 2014, https://www.traveldailynews.com/post/uae-businessman%E2%80%99s-average-use-of-private-jets-is-100-to-150-hours-per-year%2C-against-50-to-100-hours-in-europe-59791 (accessed June 7, 2020).

40. Ibid.

41. https://en.wikipedia.org/wiki/Royal_Flight_of_Oman (accessed January 24, 2019).

42. "Population of Cities in United Arab Emirates (2020)," World Population Review, http://worldpopulationreview.com/countries/united-arab-emirates-population/cities/ (accessed June 7, 2019).

43. Most expatriates living in Dubai are Asians, mainly from India and Pakistan, and tend to be employed as menial workers. European, North American and Australian expatriates in Dubai are mostly involved in the real estate, tourism, IT, oil, and healthcare sectors. https://en.wikipedia.org/wiki/Dubai#Demographics.

44. The Living Planet Report, a science-based analysis of the impact of human activity on the planet, is published every two years by the World Wildlife Fund for Nature since 1998.

45. https://en.wikipedia.org/wiki/Transportation_in_Dubai#Roads (accessed January 24, 2019).

46. A specialized authority, RTA, was established in 2005 and now operates a reasonable bus network and a newly constructed metro.

47. Pablo Paster, "Ask Pablo: Is Indoor Skiing Really That Bad?," Treehugger, December 14, 2010, https://www.treehugger.com/clean-technology/ask-pablo-is-indoor-skiing-really-that-bad.html (accessed January 24, 2019).

48. Andrew Topf, "World's 10 Biggest Energy Gluttons," *Worldpress*, October 3, 2014, https://www.worldpress.org/article.cfm/Worlds-10-Biggest-Energy-Gluttons (accessed January 24, 2019).

49. Dubai uses natural gas for electricity generation, so its power stations emit less CO_2 than would have been the case if they were using oil. The actual emissions associated with the power consumed to cool Ski Dubai is probably 25 percent less than the amount calculated for the stated number of barrels of oil and would probably amount to "only" 413,000 tons CO_2 per annum.

50. For current prices see "Gasoline Prices Around the World," Global Petrol Prices, https://www.globalpetrolprices.com/gasoline_prices/ (accessed January 24, 2019).

51. IRENA, "Renewable Energy Management Analysis: GCC 2019," 2019: 43.

52. https://www.globalpetrolprices.com/electricity_prices/ (accessed January 24, 2019).

53. Ibid.

54. Bahrain, with 19,592 Kw/h per capita per annum, ranks third in the world on that score. Qatar (15,309 kw/h) ranks seventh;

Kuwait (15,213), eighth; and UAE (11,264 kw/h per capita per annum), tenth, https://data.worldbank.org/indicator/eg.use.elec.kh.pc (accessed January 24, 2019).

55. The World Bank, *World Development Report 2003: Sustainable Development in a Dynamic World -Transforming Institutions, Growth, and Quality of Life* (New York: Oxford University Press, 2003) focused early attention on how climate hazards might intersect with multidimensional inequalities to generate uneven impacts on people and their livelihoods. The Stern report (Nicolas Stern, *The Stern Review on the Economics of Climate Change* [London: HM Treasury, 2006]) noted that climate change will deepen poverty by impacting agriculture, flooding, malnutrition, water resources, and health. The interaction of climate change and human development was analyzed in Liliana Carvajal-Velez, "Impacts of Climate Change on Human Development," Human Development Report Office Occasional Paper, New York: United Nations Development Programme, 2007; in the UN's Economic Commission for Africa (United Nations, Economic Commission for Africa, "Climate Change and Human Development Issues," Paper No. 3, 2010; and in Barry B. Hughes et. al., "Exploring Future Impacts of Environmental Constraints on Human Development," *Sustainability* 4, no. 5 [2012]: 958–994).

56. IPCC, "AR 5 Climate Change 2014: Impacts, Adaptation and Vulnerability," 2014, https://www.ipcc.ch/report/ar5/wg2/ (accessed May 24, 2019).

57. Robin Leichenko and Julie A. Silva, "Climate Change and Poverty: Vulnerability, Impacts and Alleviation Strategies," *WIREs Climate Change* 5, no. 4 (July/August, 2014): 539–556.

58. G7 Germany, "A New Climate for Peace: Taking Action on Climate and Fragility Risks. An Independent Report Commissioned by the G7 Members," 2015, https://www.newclimateforpeace.org/ (accessed May 24, 2019).

59. The marsh Bedouins of southeast Iraq and the Ghawarneh in Israel and Jordan are two obvious examples.

60. For two early reviews of the nexus between climate change and poverty, see World Bank *Global Monitoring Report 2008: MDGs and the Environment—Agenda for Inclusive and Sustainable Development* (Washington, D.C: World Bank, 2008); and Brookings Institute (Lael Brainard, Abigail Jones, and Nigel Purvis, eds.), *Climate Change and Global Poverty: A Billion Lives in the Balance?* (Washington, D.C: Brookings Institution Press, 2009).

61. Countries analyzed in terms of the climate and poverty nexus included, among others, Tanzania (Jouni Paavola, "Livelihoods, Vulnerability and Adaptation to Climate Change in the Morogoro Region, Tanzania," *Environmental Science and Policy* 11, no. 7 [2008]: 642–654); India (E. Somanathan, and Rohini Somanathan, "Climate Change: Challenges Facing India's Poor," *Economic and Political Weekly* 44, no. 31 [2009]: 51–58); and Nepal (Popular Gentle, et. al., "Differential Impacts of Climate Change on Communities in the Middle Hills Region of Nepal," *Natural Hazards* 74, no. 2 [2014]: 815–836).

62. Studies of the impact of poverty and climate change on agriculture include Syud A. Ahmed, Noah S. Diffenbaugh, and Thomas W. Hertel, "Climate Volatility Deepens Poverty Vulnerability in Developing Countries," *Environmental Research Letters* 4, no. 3 (2009): 1–8; and Thomas W. Hertel and Stephanie D. Rosch, "Climate Change, Agriculture and Poverty" World Bank Policy Research Working Paper No. 5468 (Washington D.C.: World Bank, 2010).

63. See David Satterthwaite et. al, "Adapting to Climate Change in Urban Areas: The Possibilities and Constraints in Low- and Middle-Income Nations," Human Settlements Discussion Paper (London: Human Settlements Group and Climate Change Group at the International Institute for Environment

and Development, 2007); Jorgelina Hardoy and Gustavo Pandiella, "Urban Poverty and Vulnerability to Climate Change in Latin America," *Environment and Urbanization* 21, no. 1 (2009): 203–224.

64. See for examples World Health Organization, "Improving Public Health Responses to Extreme Weather/ Heat-Waves. Summary for Policy-Makers: EuroHEAT—Technical Summary" (Copenhagen: WHO Regional Office for Europe, 2009); C. J. Portier and others, *A Human Health Perspective on Climate Change: A Report Outlining the Research Needs on the Human Health Effects of Climate Change* (Research Triangle Park, NC: Environmental Health Perspectives/National Institute of Environmental Health Sciences, 2010).

65. M. L. Parry, C. Rosenzweigb, A. Iglesiasc, M. Livermore, and G. Fischere, "Effects of Climate Change on Global Food Production under SRES Emissions and Socio-Economic Scenarios," *Global Environmental Change* 14, no. 1 (2004): 53–67.

66. Rebecca Leber, "A Hot Planet Threatens Food and Water More than Realized, IPCC Says," Grist, August 8, 2019, https://grist.org/article/a-hot-planet-threatens-food-and-water-more-than-realized-ipcc-says/ (accessed October 20, 2019).

67. Marc Sadler and Nicholas Magnan, "Grain Import Dependency in the MENA Region: Risk Management Options," Researchgate, February 2017, https://www.researchgate.net/publication/226684899_Grain_import_dependency_in_the_MENA_region_risk_management_options (accessed June 13, 2019).

68. Agora, "Food Security and Cereal Import Dependency," Agora MedSpring, January 16, 2016, http://agora.medspring.eu/en/articles/food-security-and-cereal-import-dependency. (accessed June 13, 2019).

69. Erica Denicola, Omar S. Aburizaiza, Azhar Siddique, Haider Khwaja, David Carpenter, "Climate Change and Water

Scarcity: The Case of Saudi Arabia," *Annals of Global Health* 81, no. 3 (2015): 342–353.

70. Ibid., 345.

71. Stratfor, *"A Saudi Water Crisis Lurks Beneath the Surface," Stratfor,* January 5, 2017, https://worldview.stratfor.com/article/saudi-water-crisis-lurks-beneath-surface (accessed October 31, 2018).

72. N. A Samad and V. L. Bruno, "The Urgency of Preserving Water Resources," *EnviroNews* 21 (2013): 3–6.

73. The actual amount extracted rose from 3.6 million m^3 in 1980 to 15.5 million m^3 in 2005. http://sawea.org/pdf/FutureOfSaudiArabianWaterAquifers.pdf (accessed June, 12 2019).

74. See https://www.indexmundi.com/agriculture/?country=sa&commodity=wheat&graph=exports (accessed June 13, 2019).

75. According to some estimates, natural water resources in parts of Saudi Arabia could disappear completely by 2040 (ibid.).

76. Francesca de Châtel, "The Role of Drought and Climate Change in the Syrian Uprising: Untangling the Triggers of the Revolution," *Middle Eastern Studies* 50, no. 4 (2014): 521–535.

77. Ibid.

78. Ibid.

79. Ibid.

80. "GDP—Per Capita (PPP) 2007 Country Ranks," Photius, September 15, 2007, https://photius.com/rankings/economy/gdp_per_capita_2007_0.html (accessed June 12, 2019).

81. The GDP figure pertains to purchase power parity. Source: https://data.worldbank.org/indicator/NY.GDP.PCAP.CD?locations=SY (accessed June 12, 2019).

82. See for example the battles in Allepo and at the Tishrin hydroelectric dam on the Euphrates River in 2012, mentioned

in P. H. Gleick, "Water, Drought, Climate Change, and Conflict in Syria," *Weather, Climate and Society* 6 (2014): 331–340.

83. World Economic and Social Survey, "Chapter II: Climate Change and Inequality Nexus," 2016: 21–46, https://wess.un.org/wp-content/uploads/2014/11/2_Chapter_WESS2016.pdf (accessed June 12, 2019). See also https://wess.un.org/documents/link-between-climate-change-and-inequality/ (accessed June 12, 2019).

84. Intergovernmental Panel on Climate Change, "AR5 Synthesis Report: Climate Change 2014," October 2014, https://www.ipcc.ch/assessment-report/ar5/ (accessed April 8, 2020).

85. Matt McGrath, "Climate Change: COP24 Fails to Adopt Key Scientific Report," BBC, December 8, 2018, https://www.bbc.com/news/science-environment-46496967 (accessed June 13, 2019).

86. Algeria, ranking 10th in the world in terms of landmass, is slightly ahead of Saudi Arabia, which ranks 13th, http://worldpopulationreview.com/countries/countries-in-world-by-area/ (accessed June 13, 2019).

87. ECO Newsletter, September 7, 2000. Cited in Joanna Depledge, "Striving for No: Saudi Arabia in the Climate Change Regime," *Global Environmental Politics* 8 no. 4 (2008): 11.

88. ""Ibid., 9–35.

89. Kelly Mcevers, "Saudi Arabia Tries to Stall Global Emissions Limits," NPR, December 10, 2009, https://www.npr.org/templates/story/story.php?storyId=121282056 (accessed June 14, 2019).

90. Depledge, "Striving for No," 9–35.

CHAPTER 3

1. Fabrice Balanche, *Water Issues Are Crucial to Stability in Syria's Euphrates Valley,* POLICYWATCH 2622 (Washington

D.C: Washington Institute, 2016), https://www.washingtoninstitute.org/policy-analysis/view/water-issues-are-crucial-to-stability-in-syrias-euphrates-valley (accessed June 17, 2019).

2. Ibid.

3. Ibid.

4. Ibid.

5. Ibid.

6. Ziad Awad, "Deir Al-Zor after Islamic State: Between Kurdish Self Administration and a Return of the Syrian Regime," European University Institute 2018, http://cadmus.eui.eu/bitstream/handle/1814/52824/RPR_2018_02_Eng.pdf?sequence=4&isAllowed=y(accessed June 18, 2019).

7. "Deir al-Zour: The Forgotten City," Al Ameen for Humanitarian Relief, 2017, https://reliefweb.int/sites/reliefweb.int/files/resources/Deir_Al_Zour_The_Forgotten_City.pdf (accessed June 25, 2019).

8. Ibid.

9. "Syria: Thousands of Displaced Confined to Camps," Human Rights Watch, August 1, 2018, https://www.hrw.org/news/2018/08/01/syria-thousands-displaced-confined-camps (accessed June 25, 2019).

10. Arnon Karnieli, Alexandra Shtein, Natalya Panov, Noam Weisbrod, and Alon Tal, "Was Drought Really the Trigger Behind the Syrian Civil War in 2011?" *Water* 11 (2019): 1564.

11. "UN Report of the International Commission of Inquiry on Darfur to the United Nations Secretary General," 2005, https://www.un.org/ruleoflaw/files/com_inq_darfur.pdf (accessed July 6, 2020).

12. P. Droogers et al., "Water Resources Trends in Middle East and North Africa towards 2050," *Hydrology and Earth System Sciences,* 16 (2012): 3101–3114.

13. For rural adaptation strategies to climate change see I. Sasson and A. Weinreb, "Land Cover Change and Fertility in West-Central Africa: Rural Livelihoods and the Vicious Circle Model," *Population and Environment* 38, no. 4 (2017): 345–368.

14. Quentin Wodon, Andrea Liverani, George Joseph, and Nathalie Bougnoux (eds.) *Climate Change and Migration: Evidence from the Middle East and North Africa*, chapter 7 (Washington D.C: The World Bank, 2014).

15. The survey distinguished resident migrants (people who moved away but were still considered part of their household of origin) from nonresident immigrants who, having moved permanently, were no longer counted in their family of origin (ibid.).

16. Syrians linked emigration to droughts significantly more frequently than respondents in the other four countries. Figures were mostly similar for resident and nonresident migrants. So was, for the most part, statistical significance (ibid.).

17. Ibid.

18. Ibid.

19. E. El-Hinnawi, *Environmental Refugees* (Nairobi, Kenya: United Nations Environmental Programme, 1985). As part of this definition, *environmental disruption* means "physical, chemical, and/or biological changes in the ecosystem (or resource base) that render it, temporarily or permanently, unsuitable to support human life."

20. Intergovernmental Panel on Climate Change, "Overview of Impacts, Adaptation, and Vulnerability to Climate Change," Working Group 2, Chapter 1, 2001, https://www.ipcc.ch/site/assets/uploads/2018/03/WGII_TAR_full_report-2.pdf.

21. UNHCR, "Climate Change, Natural Disasters and Human Displacement: A UNHCR Perspective," n.d., http://www.unhcr.org/protection/environment/4901e81a4/unhcr-policy-paper-climate-change-natural-disasters-human-displacement.html (accessed July 15, 2019). Footnote 1 states that "This policy paper updates and supersedes the paper issued by UNHCR in October 2008 and will be updated as necessary, in line with relevant developments within the climate change debate."

22. The rather confusing list included hydro-meteorological disasters (flooding, typhoons, mudslides, etc.); instances by which governments formally designate "zones of high risk"; environmental degradation and slow disasters such as water shortage, desertification, and salinization; submergence of small island states; and wars over resources such as water, land, and food that stem from shortages created by climatic change.

23. In certain circumstances, the UNHCR report conceded, people displaced from islands that become submerged might also qualify.

24. See for example Frank Bierman and Ingrid Boas, "Protecting Climate Refugees: The Case for a Global Protocol," *Environment: Science and Policy for Sustainable Development*, November-December 2008, https://www.tandfonline.com/doi/abs/10.3200/ENVT.50.6.8–17 (accessed July 16, 2019).

25. UNHCR, "Climate Change, Natural Disasters and Human Displacement." https://www.unhcr.org/4901e81a4.pdf

26. The political base of the Houthi rebels is premised on restructuring relations between tribes in the north, uniting them against the Saudi-backed Saleh regime in the south.

27. "Government Effectiveness—Country Rankings (2018)," The Global Economy.com, https://www.theglobaleconomy.com/rankings/wb_government_effectiveness/ (accessed July 28, 2019).

28. Ibid. The global index (193 countries) ranges from the highest score of 2.21 (Singapore) to the lowest of –2.21 (Somalia). U.A.E (1.4), Israel, Qatar, Saudi Arabia, Oman, Bahrain, Jordan, and Turkey score above zero. All other MENA countries score below zero with Sudan (–1.41, ranked 180), Libya (–1.77, ranked 189), Syria (–1.79, ranked 190), and Yemen (–192, ranked 191) scoring particularly low.

29. Ibid. Libya (ranked 189), Syria (190), Yemen (191), and Somalia (193). Iraq (ranked 179) and Sudan (180) are not far higher on the scale.

30. "Political Stability—Country Rankings (2018)," The Global Economy.com, https://www.theglobaleconomy.com/rankings/wb_political_stability/ (accessed July 28, 2019).

31. Barry Buzan, Ole Wæver, and Jaap de Wilde, *Security: A New Framework for Analysis* (Boulder, CO: Lynne Rienner, 1998).

32. See Peter Hough, *Understanding Global Security* (London: Routledge, 2004); Barry Buzan, *People, States, and Fear: The National Security Problem in International Relations* (Brighton, Sussex: Wheatsheaf, 1983); William Bain, *The Empire of Security and the Safety of the People* (London: Routledge, 2006).

33. Betsy Hartman, "Rethinking Climate Refugees and Climate Conflict: Rhetoric, Reality and the Politics of Policy Discourse," *Journal of International Development* 22 (2010): 233–246. See also Maria Julia Trombetta, "Linking Climate-Induced Migration and Security within the EU: Insights from the Securitization Debate," *Critical Studies on Security* 2, no. 2 (2014): 131–147.

34. After Gilles Deleuze and Felix Guattari, *A Thousand Plateaus* (Minneapolis: University of Minnesota Press, 1987).

35. Gregory White, *Climate Change and Migration: Security and Borders in a Warming World* (Oxford: Oxford University Press, 2011).

36. For a critical assessment of this logic see: Romain Felli, "Managing Climate Insecurity by Ensuring Continuous Capital Accumulation: 'Climate Refugees' and 'Climate Migrants,'" *New Political Economy* 18, no. 3 (2012): 337–363.

37. Giovanni Bettini, "Climate Migration as an Adaption Strategy: De-Securitizing Climate-Induced Migration or Making the Unruly Governable?" *Critical Studies on Security*, 2, no. 2 (2014): 180–195.

38. Frank Bierman and Ingrid Boas, "Protecting Climate Refugees: The Case for a Global Protocol," *Environment* 50, no. 6 (2010): 10–16; Frank Bierman and Ingrid Boas, "Climate Change and Human Migration: Towards a Global Governance System to Protect Climate Refugees," in *Climate Change, Human Security and Violent Conflict: Challenges for Societal Stability*, eds. Jürgen Scheffran, Michael Brzoska, Hans Günter Brauch, Michael Link, and Janpeter Schilling (New York: Springer, 2012).

39. Ibid.

40. K. Warner, M. Hamza, A. Oliver-Smith, F. Renaud, and A. Julca, "Climate Change, Environmental Degradation and Migration," Natural Hazards 55 (2010): 689–715, https://www.researchgate.net/profile/Alex_Julca/publication/225452578_Climate_change_environmental_degradation_migration/links/54e3640f0cf2748d3a9d6387.pdf (accessed July 25, 2019).

CHAPTER 4

1. J. C. Jones et al., "Total Amounts of Oil Produced over the History of the Industry," *International Journal of Oil Gas and Coal Technology*, 2, no. 2 (2009): 199.

2. The Middle East currently accounts for 40 percent of global world production and approximately half of known reserves. In earlier decades, before Russia and the US upped their share of overall global production, the Middle East's proportional weight was higher than it is today.

3. The misalignment between the overall amount of oil and gas historically produced in the Middle East (about half of all the oil ever produced globally) and the estimated contribution of this amount to historic greenhouse gas emissions (about a third of the global amount) has to do with coal. Burning coal produces more than twice the CO_2 emitted from burning oil.

Since the Middle East has never been a source of coal, its proportional contribution to historic CO_2 emissions is lower than its share in the historic global extraction of oil.

4. Rym Ghazal, "When Abu-Dhabi Had 30 Cars," The National, AE, November 3, 2011, https://www.thenational.ae/uae/when-abu-dhabi-had-30-cars-1.432812 (accessed September 3, 2019).

5. Shafaat Shahbandari, "For Every Two Dubai Residents, There Is One Car," Gulf News, March 15, 2015, https://gulfnews.com/uae/transport/for-every-two-dubai-residents-there-is-one-car-1.1472177 (accessed September 3, 2019).

6. Liz Cookman, "More than Three Million Cars on UAE Roads, Report Finds," The National, AE, December 9, 2018, https://www.thenational.ae/uae/more-than-three-million-cars-on-uae-roads-report-finds-1.800907 (accessed September 3, 2019).

7. "Kuwait Energy Production and Consumption," Ceicdata.com, https://www.ceicdata.com/en/kuwait/energy-production-and-consumption (accessed September 2019).

8. The average per capita electricity consumption across the US in 1973 was 8,573 kw/h, https://www.google.com/search?q=US+electricity+consumption+per+capita+1973&oq=US+electricity+consumption+per+capita+1973&aqs=chrome.69i57.9023j0j4&sourceid=chrome&ie=UTF-8 (accessed September 3, 2019); by 2017 it grew to 13,005 kw/h, https://en.wikipedia.org/wiki/Energy_in_the_United_States (accessed September 3, 2019).

9. https://en.wikipedia.org/wiki/OPEC#cite_note-qtrleave-122 (accessed April 13, 2020).

10. https://en.wikipedia.org/wiki/Gulf_Cooperation_Council (accessed April 13, 2020).

11. Saudi Arabia is often cited as OPEC's de facto leader. Kuwait and the UAE are active members, as was Qatar until 2019 when it removed itself and joined Oman, Bahrain, Russia, and seven other oil producing countries in OPEC Plus https://en.wikipedia.org/wiki/OPEC (accessed April 3, 2020).

OPEC and OPEC Plus normally attempt to coordinate production quotas and pricing, although in early 2020 a major split between them, which came as a result of Saudi Arabia and Russia pulling in opposite directions, had detrimental consequences for members of both institutions.

12. In 1973 the UAE's GDP was just under $3 billion. In 2018 it reached $414 billion, https://tradingeconomics.com/united-arab-emirates/gdp (accessed September 2, 2019).

13. "Saudi Arabia GDP—Gross Domestic Product," Country Economy.com, https://countryeconomy.com/gdp/saudi-arabia?year=1973 (accessed September 2, 2019).

14. "USA GDP—Gross Domestic Product," CountryEconomy.com, https://countryeconomy.com/gdp/usa?year=1973 (accessed September 2, 2019).

15. Ibid.

16. See J. Lelieveld, Y. Proestos, P. Hadjinicolaou et al., "Strongly Increasing Heat Extremes in the Middle East and North Africa (MENA) in the 21, Century," *Climatic Change 137*, no. 1-2: 245–260; See also the 2016 report by Max Planck Institute entitled "Climate Exodus Expected in the Middle East and Africa," https://www.mpg.de/10481936/climate-change-middle-east-north-africa (Accessed October 18, 2018).

17. IRENA, "Renewable Energy Market Analysis: The GCC Region" (Abu Dhabai: IRENA, 2019), Fig. 1.4, Fig. 11.

18. Sörgel clearly saw his plan as a benign, more equitable alternative to other visions Germans had for the Middle East and Africa which later formed the backbone of the Nazi colonizing drive in the Balkans, Southern Europe, and beyond.

19. Quora Contributor," We Could Power the Entire World by Harnessing Solar Energy from 1% of the Sahara," *Forbes*, September 22, 2016, https://www.forbes.com/sites/quora/2016/09/22/we-could-power-the-entire-world-by-harnessing-solar-energy-from-1-of-the-sahara/#456e2ec6d440 (accessed May 19, 2019).

20. For an overview of this vision see desertec.org. (accessed June 9, 2020). While Desertec's vision does not currently seem economically viable, the energy calculation informing it remains valid.

21. "The World Is Changing—Impact Report 2019," Bloomberg.com, https://data.bloomberglp.com/company/sites/56/2020/05/Impact-Report-2019d.pdf (accessed June 9, 2020).

22. https://en.m.wikipedia.org/wiki/Electricity_sector_in_China (accessed August 12, 2019).

23. https://en.wikipedia.org/wiki/Solar_power_in_Italy (accessed August 12, 2019).

24. https://en.wikipedia.org/wiki/List_of_countries_by_electricity_production_from_renewable_sources

25. Justin Walton, "The 5 Countries That Produce the Most Solar Energy," Investopedia, June 25, 2019, https://www.investopedia.com/articles/investing/092815/5-countries-produce-most-solar-energy.asp (accessed August 12, 2019).

26. Ibid.

27. Yuki Yu, "China Takes New Road to Subsidy-Free Offshore Wind," Recharge News, September 17, 2018, https://www.rechargenews.com/wind/1559071/china-takes-new-road-to-subsidy-free-offshore-wind (accessed August 15, 2019).

28. A gigawatt is 1,000 megawatts. A country with 10 million inhabitants would normally have a generating capacity of between 10 and 20 gigawatts.

29. "Late Surge in Offshore Wind Financings Helps 2019 Renewables Investment to Overtake 2018," Bloomberg NEF, January 16, 2020, https://about.bnef.com/blog/late-surge-in-offshore-wind-financings-helps-2019-renewables-investment-to-overtake-2018/ (accessed March 7, 2020).

30. See for one example Atlantica, an energy company with a turnover of over half a billion dollars in 2019. Its earnings from renewables in the fourth quarter of 2019 was $111 million, 12.1 percent higher than the figure for the fourth quarter of 2018. See Marco Di Lallo, "This High-Yielding Renewable-

Energy Stock Sees Another Year of High-Powered Growth in 2020," Motley Fool, March 4, 2020, https://www.fool.com/investing/2020/03/04/this-high-yielding-renewable-energy-stock-sees-ano.aspx (accessed March 6, 2020).

31. The report indicates deals in the first six weeks of 2020 involving more than 6 gigawatts of solar energy, which could bring the figure for the whole quarter to over 12 gigawatts. The entire first quarter of 2019, in comparison, saw solar deals signed involving only 10.5 gigawatts. See John Parnell, "Behind the Flurry of Big Solar Deals in 2020: Conventional Energy Companies," Greentech Media.com, February 11, 2020, https://www.greentechmedia.com/articles/read/oil-majors-and-utilities-drive-flurry-of-solar-deals-in-2020 (accessed April 3, 2020).

32. "142 GW of Global Solar Capacity Will Be Added in 2020, Says HIS," Renewable Energy World, October 1, 2020, https://www.renewableenergyworld.com/2020/01/10/142-gw-of-solar-capacity-will-be-added-to-the-global-market-in-2020-says-ihs/ (accessed March 7, 2020).

33. "Chart: Renewable Energy Costs Are Falling," Inside Climate News, December 1, 2018, https://insideclimatenews.org/content/chart-renewable-energy-costs-are-falling (accessed March 14, 2020).

34. Ibid.

35. The terawatt, the equivalent of 1,000 gigawatts, is a measure used mostly for estimates of exceptionally large quantities of electricity. For example, the total electricity consumption of the US in 2017 was 1.2 terawatts.

36. New solar installations between 2009 and 2019 accounted for 0.683 terawatts; wind turbines accounted for 0.487 terawatts, and hydro for 0.283 terawatts. See William Mathis, "Clean Energy Investment Is Set to Hit $2.6 Trillion This Decade," Bloomberg, September 5, 2019, https://www.bloomberg.com/news/articles/2019–09–05/clean-energy-investment-is-set-to-hit-2–6-trillion-this-decade (accessed March 15, 2020).

37. The balance is expected to be filled by nuclear and by other renewables. See Renewable Energy World (2020), https://www.renewableenergyworld.com/2020/01/10/142-gw-of-solar-capacity-will-be-added-to-the-global-market-in-2020-says-ihs/

38. This represents a 1,100 percent growth in 32 years. By Bloomberg's analysis (see note 36) wind power generation, currently providing 5 percent of global energy, is destined to grow by 2050 to 26 percent; hydroelectric will remain constrained by resource availability and thus see modest growth; while nuclear, considerably more expensive and lacking the flexibility required if it is to complement renewables in smart grids, will stay at roughly current levels.

39. https://www.greentechmedia.com/articles/read/california-renewable-curtailments-spike-as-coronavirus-reduces-demand

40. Ibid.

41. For a prognosis of how Covid-19 could propel intensified transition to renewable energy see https://www.theguardian.com/environment/2020/may/19/how-renewable-energy-could-power-britains-economic-recovery.

42. A megawatt is equivalent to a million watts. A light bulb requires 50 to 100 watts. A single coil heater consumes 750 to 1,000 watts, a measure also known as 1 kilowatt. A megawatt (1,000 kilowatts) can keep a few hundred households going simultaneously. A power plant feeding a mid-sized town needs to generate some 1,000 megawatts.

43. Michael Hochberg, "Renewable Energy Growth in Morocco: An Example for the Region," Middle East Institute, March 2016, https://www.mei.edu/sites/default/files/publications/PF26_Hochberg_Moroccorenewables_web.pdf (accessed August 14, 2019).

44. https://en.wikipedia.org/wiki/Ouarzazate_Solar_Power_Station#Noor_I (accessed August 14, 2019). Noor is a concentrated solar power (CSP) plant, where large-scale lens and mirrors focus sunlight to create high energy. For more on

this and on the difference between CSP and the more prevalent photovoltaic installations see below.

45. https://en.wikipedia.org/wiki/Renewable_energy_in_Turkey (accessed August 14, 2019).

46. International Trade Administration, "Egypt, Renewable Energy: This Is a Best Prospect Industry Sector for This Country," Trade.gov, October, 13, 2019, https://www.export.gov/article?id=Egypt-Renewable-Energy (accessed June 7, 2020).

47. Ibid.

48. Pouyan Jan, "Renewables Are Booming in Iran," Medium, December 26, 2018, https://medium.com/age-of-awareness/renewables-are-booming-in-iran-98781650669d (accessed August 12, 2019).

49. Ibid.

50. Emiliano Bellini, "Iraq Takes First Concrete Steps into Solar," PV Magazine, January 8, 2019, https://www.pv-magazine.com/2019/01/08/iraq-takes-first-concrete-steps-into-solar/ (accessed August 14, 2019).

51. IRENA, "Renewable Energy Market Analysis: The GCC Region" (Abu Dhbai: IRENA, 2019), 14.

52. Ibid, 50.

53. John Parnell, "World's Largest Solar Power Plant Switched On," *Forbes*, June 29, 2019, https://www.forbes.com/sites/johnparnell/2019/06/29/worlds-largest-solar-power-plant-switched-on/#269d314d161a (accessed August 12, 2019).

54. "United Arab Emirates Renewable Energy," International Trade Administration, October 13, 2019, https://www.trade.gov/knowledge-product/united-arab-emirates-renewable-energy (accessed June 7, 2020).

55. Ibid., 58–59.

56. Ibid.

57. The technology selected for this phase was concentrated solar power (CSP).

58. See note 54, page 59.

59. E. Bellini, "SirajPower Develops 50 MW under DEWA's Shams Dubai Initiative," PV Magazine, October 10, 2018, https://www.pv-magazine.com/2018/10/10/sirajpower-develops-50-mw-under-dewas-shams-dubai-initiative/ (accessed August 11, 2019).

60. Parnell, "World's Largest Solar Power Plant Switched On."

61. See Parnell, note 53.

62. "The Plan to Turn Saudi Arabia into a Renewable Energy Leader," Oxford Business Group, April 26, 2019, https://oxfordbusinessgroup.com/news/plan-turn-saudi-arabia-renewable-energy-leader (accessed July 31, 2019).

63. IRENA, "Renewable Energy Market Analysis: The GCC Region" (Abu Dhabi: IRENA, 2016).

64. IRENA, "Renewable Energy Market Analysis," 2019, 59–60.

65. Farid Rahman, "Saudi Arabia Plans $1.51b Investment in Renewable Energy," Gulf News, January 29, 2019, https://gulfnews.com/business/energy/saudi-arabia-plans-151b-investment-in-renewable-energy-1.61742719 (accessed August 12, 2019).

66. REPDO Renewable Energy Project Development Office, "The National Renewable Energy Program," 2018, www.powersaudiarabia.com.sa/web/index.html (accessed August 13, 2019).

67. S. Cunningham and V. Nereim, "Saudis, SoftBank Plan World's Largest Solar Project," Bloomberg, March 28, 2018, www.bloomberg.com/news/articles/2018–03–28/saudi-arabia-softbank-ink-deal-on-200-billion-solar-project (accessed August 13, 2019). Unconfirmed reports later in 2018 however suggested that the project was suspended (IRENA, "Renewable Energy Market Analysis," 2019, 60).

68. Government of Saudi Arabia, "Neom," 2018, www.neom.com (accessed August 13, 2019).

69. See Rahman, note 65.

70. Riyadh- Fateh al-Rahman Youssef, "Saudi Arabia Plans to Invest $80 Bn in Renewable Energy," Asharq Al-Awsat, March 5, 2019, https://aawsat.com/english/home/article/1619766/saudi-arabia-plans-invest-80-bn-renewable-energy (accessed August 12, 2019). For comparison, according to Bloomberg's 2018 global renewable investment report (accessed August 10, 2019), the five world leaders in investment in renewable energy in 2018 were China ($100.1 billion), US ($64.2 billion), Japan ($27.2 billion), India ($11.1 billion), and Germany ($10.5 billion).

71. "Saudi Arabia to Become Largest Middle East Wind Power Market by Early 2020s," Wood Mackenzie, April 8, 2019, https://www.woodmac.com/press-releases/saudi-arabia-to-become-largest-middle-east-wind-power-market-by-early-2020s/ (accessed August 14, 2019).

72. IRENA, "Renewable Energy Market Analysis," 2019.

73. Ibid., 56–58.

74. Other GCC territories are finally awakening to electricity pricing reforms. The UAE was first to remove electricity subsidies for expatriates and, since 2018, has imposed a 5 percent value added tax on fuel. In Abu Dhabi in 2017 electricity rates increased by 34 percent for UAE nationals and 28 percent for expatriates. Qatar increased electricity rates in 2015 as did Oman in 2017 for about 10,000 customers consuming more than 150 MWh per year. Bahrain was about to increase its electricity tariff for industrial, commercial, and most residential customers to 7.7 US cents/kWh by late 2019 (IRENA, Renewable Energy Market Analysis," 2019, 124).

75. Oxford Business Group, "The Plan to Turn Saudi Arabia into a Renewable Energy Leader" (see note 62).

76. For comparison, in 2017 Saudi Arabia's subsidized national power tariff—the price of electricity of all kinds collected from end consumers—fluctuated between 1.33 US cents for initial consumption and 7 US cents for heavy users. See John Parnell,

"Saudi Arabia Reaches Renewable Energy Crossroads in 2019," Forbes, December 27, 2018, https://www.forbes.com/sites/johnparnell/2018/12/27/saudi-arabia-reaches-renewable-energy-crossroads-in-2019/#3c3580b3738b (accessed August 15, 2019).

77. See Sagatom Saha And David Livingston, "Saudi Arabia's Renewables Revolution: How Washington Can Guide the Process," Foreign Affairs, June 6, 2017, https://carnegieendowment.org/2017/06/06/saudi-arabia-s-renewables-revolution-pub-71183 (accessed August 15, 2019).

78. Extrapolated from https://en.wikipedia.org/wiki/List_of_cities_by_sunshine_duration. (accessed May 20, 2019). Marsa Alam in eastern Egypt, for example, has as a yearly average 10.84 hours per day of sunshine—almost 90 percent of daytime hours. Khartoum scores 85 percent on that scale; Muscat (in Oman) and Dubai, 80 percent; Ourazette (Morocco), 78 percent; Tel-Aviv, 76 percent; Gabes, (Tunisia) 75 percent; Baghdad and Riad, 74 percent; to name but a few.

79. For figures of the global average, see https://en.wikipedia.org/wiki/Sunlight (accessed August 5, 2019). For an overall estimate of the solar potential of the Sahara desert, see http://www.desertec.org/ (accessed June 7, 2019).

80. The relevant index for the suitability of a location for PV is its global horizontal irradiance (GHI), which measures both direct sunlight reaching the surface and indirect light diffused by floating particles.

81. The suitability of a location for CSP is a function of the direct normal irradiation (DNI) measured over time. The DNI figure gauges the amount of radiation that reaches a surface positioned perpendicular to the direction of the rays. Energy deflected by airborne particles is lost for CSP and is therefore subtracted from the measure.

82. Sand and dust storms can, in extreme cases, reduce direct irradiation by up to 90 percent. P. G. Kosmopoulos, "Dust Impact on Surface Solar Irradiance Assessed with Model

Simulations, Satellite Observations and Ground-Based Measurements," 2017, www.atmos-meas-tech.net/10/2435/2017/amt-10-2435-2017.pdf (accessed August 9, 2019).

83. Small amounts of energy from nuclear and smaller scale waste-to-energy and tidal wave technologies, IRENA, "Renewable Energy Market Analysis," 2019, 51.

84. Egypt's Gulf of Suez and the Jordanian highland have particularly favorable conditions for wind turbines. Parts of Saudi Arabia, Kuwait, and Oman, while not as favorable, still have average wind speeds that are higher than the required 7.5 meters per second at 100 meters altitude (ibid., 33).

85. https://en.wikipedia.org/wiki/Land_use_statistics_by_country (accessed August 5, 2019).

86. This rough figure is representative of the region's five largest countries Algeria, Saudi Arabia, Sudan, Libya, and Egypt which among them account for two-thirds of the region's landmass. Algeria, with 2.38 million square kilometers, has 3.5 percent of its territory cultivated. Saudi Arabia, with 2.15 million km^2, has 1.6 percent farmed. Sudan (1.86 million km^2), 15.9 percent; Libya (1.76 million km^2), 1.2 percent; Egypt (1 million km^2), 3.6 percent; and Yemen (0.52 million km^2), 1.8 percent (ibid.).

87. "The World is Changing: Bloomberg Impact Report 2019," Bloomberg, https://data.bloomberglp.com/company/sites/56/2020/05/Impact-Report-2019d.pdf (accessed June 7, 2020). Of the overall amount of $13 trillion needed for replenishing the global electricity supply, wind is estimated to attract $5.3 trillion, solar installations will draw $4.2 trillion, new fossil fuel plants $2 trillion, and batteries $843 billion. The total sum that needs to be invested annually to cover new installations of both renewable and conventional power production is approximately $416 billion.

88. "World Energy Investment 2018," International Energy Agency, July 2018, https://www.iea.org/wei2018/ (accessed August 7, 2019). Investment in renewables that year was more than double the $132 billion invested in power plants which burn fossil

fuel (ibid.). Frankfurt School-UNEP Centre/BNEF, "Global Trends in Renewable Energy Investment 2018," Iberglobal.com, http://www.iberglobal.com/files/2018/renewable_trends.pdf (accessed August 7, 2019), gives slightly different figures. It suggests that 2017 saw $280 billion invested in renewables, 2 percent higher than 2016 but still shy of the record $323 billion of 2015.

89. https://en.wikipedia.org/wiki/List_of_countries_by_GDP_(nominal) (accessed April 6, 2020).

90. IRENA, "Renewable Energy Market Analysis," 2019, 80.

91. Ibid.

92. Between 2010 and 2017 costs of installing PV solar installations went down by 73 percent and costs of wind turbines were cut by 22 percent. IRENA, "Renewable Energy Market Analysis," 2019, 84–85.

93. Ingrid Andersson and Thomas Andersson, "On the Role of Social Innovation in the Gulf Cooperation Council Countries," n.d., https://www.socialinnovationatlas.net/fileadmin/PDF/einzeln/02_SI-in-World-Regions/02_20_On-the-Role-of-SI-in-the-Gulf-Cooperation-Council-Countries_Andersson-Andersson.pdf (accessed July 29, 2019).

94. Farhan Aljawareen, "Innovation in the GCC Countries: An Economic Analysis," *Journal of Economics and Development Studies* 5, no. 4 (2017): 51–62; N. M. Hassan, '"Increasing the Pace towards an Innovation Economy in the Arab Region,"' *International Journal of Innovation and Knowledge Management in the Middle East and North Africa* 5, no. 2: 61–70; Omar Bizri, *Science, Technology, Innovation, and Development in the Arab Countries,* Chapter 3 (London: Academic Press, 2018).

95. See for one example Global Innovation Index (GII): 2019, gii-full-report-2019%20(1).pdf (accessed August 9, 2019).

96. Sorbonne University Abu Dhabi, "Machine Learning and Artificial Intelligence," Sorbonne University, https://www.sorbonne.ae/executive-education/machine-learning-and-artificial-intelligence/ (accessed March 25, 2020).

97. Ben Dickson, "How Qatar Plans to Become a Global Leader in Artificial Intelligence," BDTechtalks, July 24, 2019, https://bdtechtalks.com/2019/07/24/qatar-artificial-intelligence-strategy/ (accessed March 25, 2020).

98. This estimate is based on figures for combined consumption of oil in 2035 assuming renewables have not been introduced. It sees ten countries integrated to a supergrid spread across the Arabian peninsula and neighboring territories: Saudi Arabia, UAE, Kuwait, Qatar, Oman, Bahrain, Yemen, Egypt, Iraq, and Jordan. Figures for their putative oil consumption in 2035 were extrapolated from https://en.wikipedia.org/wiki/List_of_countries_by_oil_consumption (accessed August 15, 2019). The extrapolation assumed that energy consumption needs of 2011 will have been trebled by 2035.

99. According to the EPA, burning a barrel of oil emits 0.43 tons of CO_2, https://www.epa.gov/energy/greenhouse-gases-equivalencies-calculator-calculations-and-references (accessed June 9, 2020).

100. "About 7% of Fossil Fuels Are Consumed for Non-Combustion Use in the United States," US Energy Information Administration, April 6, 2018, https://www.eia.gov/todayinenergy/detail.php?id=35672 (accessed August 15, 2019). See also "Other Uses of Petroleum," Petroleum.Co.UK, http://www.petroleum.co.uk/other-uses-of-petroleum (accessed August 15, 2019).

101. Ibid.

CHAPTER 5

1. U.S.-Saudi Arabian Business Council, "The Petrochemicals and Plastics Sector in the Kingdom of Saudi Arabia," 2009, https://web.archive.org/web/20141212225752/http://www.us-sabc.org/files/public/Petrochemicals_Brochure.pdf (accessed April 1, 2020).

2. US Department of State, "Saudi Arabia– Background Notes," 2001, https://2009-2017.state.gov/outofdate/bgn/saudi arabia/26136.htm (accessed April 1, 2020).

3. Kingdom of Saudi Arabia, "Vision 2030," https://vision 2030.gov.sa/en (accessed April 1, 2020).

4. Tom Kenning, "Saudi Arabia Plans 60GW of Renewable Energy by 2030," PVTech.org, January 14, 2019, https://www.pv-tech.org/news/saudi-arabia-plans-60gw-of-renewable-energy-by-2030 (accessed April 2, 2020).

5. F. Al Harbi and D. Csala, "Saudi Arabia's Electricity: Energy Supply and Demand Future Challenges," 2019 1st Global Power, Energy and Communication Conference (GPE-COM), Nevsehir, Turkey, 2019, pp. 467–472, July 29, 2019, https://ieeexplore-ieee-org.rproxy.tau.ac.il/document/8778554 (accessed April 2, 2020).

6. UAE, "Vision 2021," updated January 19, 2020, https://u.ae/en/about-the-uae/strategies-initiatives-and-awards/federal-governments-strategies-and-plans/vision-2021 (accessed April 1, 2020).

7. UAE, "UAE Centennial 2071," updated January 16, 2020, https://u.ae/en/about-the-uae/strategies-initiatives-and-awards/federal-governments-strategies-and-plans/uae-centennial-2071 (accessed April 1, 2020).

8. UAE Government, "UAE Energy Strategy 2050", 2018, https://government.ae/en/about-the-uae/strategies-initiatives-and-awards/federal-governments-strategies-and-plans/uae-energy-strategy-2050 (accessed August 12, 2019).

9. For a critical analysis see Sophie Olver-Ellis, "Building the New Kuwait: Vision 2035 and the Challenge of Diversification," LSE Middle East Centre Paper Series (30), LSE Middle East Centre, London, http://eprints.lse.ac.uk/103198/ (accessed April 2, 2020).

10. Hukoomy, " Qatar National Vision 2030," https://portal.www.gov.qa/wps/portal/topics/Employment+and+Workplace/Qatar+National+Vision+2030 (accessed April 2, 2020).

11. Adel Peters, "This Bill Gates-Backed Solar Tech Could Help Finally Decarbonize Heavy Industry," Fast Company, https://www.fastcompany.com/90431869/this-bill-gates-backed-solar-tech-could-help-finally-decarbonize-heavy-industry (accessed April 11, 2020).

12. BBC, "Saudi Oil Attacks: Drones and Missiles Launched from Iran – US," BBC.com, September 17, 2019, https://www.bbc.com/news/world-middle-east-49733558 (accessed March 31, 2020).

13. For a comprehensive review of the role of salt in history, see Mark Kurlansky, *Salt: A World History* (New York: Penguin Books, 2003).

14. Ibid.

15. Michael Liebreich, "Covid-19—The Low-Carbon Crisis," Bloomberg NEF, March 26, 2020, https://about.bnef.com/blog/covid-19-the-low-carbon-crisis/ (accessed April 22, 2020).

16. Abigail Ng, "5 charts That Explain the Saudi Arabia-Russia Oil Price War So Far," CNBC.com, April 1, 2020, https://www.cnbc.com/2020/04/01/5-charts-that-explain-the-saudi-arabia-russia-oil-price-war-so-far.html (accessed April 22, 2020).

17. For one vivid example, China's six largest power stations burned 40 percent less coal in the first quarter of 2020 compared to the last quarter of 2019, See Ben Geman, "The Environmental Impact of China's Coronavirus Shutdown," Axios, March 18, 2020, https://www.axios.com/coronavirus-china-carbon-emissions-3453d9a1-1ae9-4789-8a41-3ed257946dbd.html (accessed April 22, 2020). See also an assessment by Michael Liebreich at BloombergNEF in late March 2020 of

the decline already recorded in energy demand as a result of the pandemic, https://about.bnef.com/blog/covid-19-the-low-carbon-crisis/ (accessed April 22, 2020).

18. Justin Worlamd, "How Coronavirus Could Set Back the Fight Against Climate Change," *Time,* March 10, 2020, https://time.com/5795150/coronavirus-climate-change/ (accessed April 18, 2020).

19. Victor Mallet, "EU Carbon Emissions Tumble During Lockdowns," *Financial Times,* April 8, 2020, https://www.ft.com/content/4c59fd16-6020-4798-b8f1-5df686bbd97a (accessed April 18, 2020).

20. Matt McGrath, "Coronavirus: Air Pollution and CO_2 Fall Rapidly as Virus Spreads," *BBC News,* March 19, 2020, https://www.bbc.com/news/science-environment-51944780 (accessed April 18, 2020).

21. Simon Evans, "Analysis: Coronavirus Set to Cause Largest Ever Annual Fall in CO_2 Emissions," Carbonbrief, April 9, 2020, https://www.carbonbrief.org/analysis-coronavirus-set-to-cause-largest-ever-annual-fall-in-co2-emissions (accessed June 10, 2020).

22. Madeleine Stone, "Carbon Emissions Are Falling Sharply Due to Coronavirus. But Not for Long," *National Geographic,* April 3, 2020, https://www.nationalgeographic.com/science/2020/04/coronavirus-causing-carbon-emissions-to-fall-but-not-for-long/ (accessed April 20, 2020).

23. Following the global financial crisis of 2008, CO_2 emissions fell 1.4 percent in 2009 but leaped to 5.9 percent the following year. Alice Hill, "What the Coronavirus Pandemic Teaches Us about Fighting Climate Change," Council on Foreign Relations, April 10, 2020, https://www.cfr.org/in-brief/what-coronavirus-pandemic-teaches-us-about-fighting-climate-change (accessed April 18, 2020).

24. To trace the price of oil over time see "Commodities," Markets Business Insider, https://markets.businessinsider.com/commodities/oil-price, (accessed June 9, 2020). For fluctuations in the spring of 2020 see BBC, "Coronavirus: Oil Price Collapses to Lowest Level for 18 Years," BBC, March 30, 2020, https://www.bbc.com/news/business-52089127 (accessed April 20, 2020).

25. Damian Carrington, Jillian Ambrose, and Matthew Taylor, "Will the Coronavirus Kill the Oil Industry and Help Save the Climate?," *The Guardian,* April 1, 2020, https://www.theguardian.com/environment/2020/apr/01/the-fossil-fuel-industry-is-broken-will-a-cleaner-climate-be-the-result (accessed April 22, 2020); https://www.bbc.com/news/business-52367052 (accessed April 22, 2020).

26. The International Renewable Energy Agency (IRENA) issued a report in early April 2020 which indicates that renewable sources now account for a third of worldwide energy production, https://www.irena.org/newsroom/pressreleases/2019/Apr/Renewable-Energy-Now-Accounts-for-a-Third-of-Global-Power-Capacity (accessed April 22, 2020).

27. Ed Crooks, "Five Ways the Coronavirus Changed the World of Energy This Week," Wood Mackenzie, March 20, 2020, https://www.woodmac.com/news/opinion/five-ways-the-coronavirus-changed-the-world-of-energy-this-week/ (accessed April 22, 2020).

28. Carrington et al., "Will the Coronavirus Kill the Oil Industry and Help Save the Climate?"

29. Resilience, "US Oil Exploration Drops by 95 Percent," 2019, https://www.resilience.org/stories/2019–05–03/us-oil-exploration-drops-by-95-percent/ (accessed April 1, 2020); MSNBC, "Quest for New Oil Discoveries Still on Back Burner," 2018, https://www.cnbc.com/2018/01/04/oil-companies-still-hesitant-over-oil-exploration.html (accessed April 1, 2020).

30. Alister Doyle, "Four More EU Nations Back a Green Post-Coronavirus Recovery," Climate Change News, April 20, 2020, https://www.climatechangenews.com/2020/04/20/four-eu-nations-back-green-post-coronavirus-recovery/ (accessed April 22, 2020).

31. Alison Tahmizian Meuse, "Saudi Dips into Reserves for Stock-Buying Spree," *Asia Times*, June 4, 2020, https://asiatimes.com/2020/06/saudi-dips-into-reserves-for-stock-buying-spree/ (accessed June 10, 2020)

32. World Bank Group, "Saudi Arabia," Climate Change Knowledge Portal, https://climateknowledgeportal.worldbank.org/country/saudi-arabia (accessed March 26, 2020).

33. Fahad Alkolibi, "Possible Effects of Global Warming on Agriculture and Water Resources in Saudi Arabia: Impacts and Responses," *Climatic Change* 54, no. 1 (2002): 225–45; Erica Denicola, Omar S. Aburizaiza, Azhar Siddique, Haider Khwaja, and David O. Carpenter, "Climate Change and Water Scarcity: The Case of Saudi Arabia," *Annals of Global Health* 81, no. 3 (2015): 342–53.

34. World Bank Group, "Kuwait," Climate Change Knowledge Portal, https://climateknowledgeportal.worldbank.org/country/kuwait (accessed March 26, 2020).

35. Manar Fawzi Bani F. Mfarrej, "Climate Change Patterns in the UAE: A Qualitative Research and Review," *Nature Environment and Pollution Technology* 18, no. 1 (2019): 261–68.

36. World Bank Group, "Qatar," Climate Change Knowledge Portal, https://climateknowledgeportal.worldbank.org/country/qatar (accessed March 26, 2020).

37. IRENA, "Renewable Energy Market Analysis: The GCC Region" (Abu Dhabai: IRENA, 2019), Fig. 1.4, Fig. 11.

38. "Qatar Quit OPEC Because of Politics, Not Oil," Economist, December 8, 2018, https://www.economist.com/middle-east-and-africa/2018/12/08/qatar-quit-opec-because-of-politics-not-oil (accessed April 13, 2020).

39. In late 2019 there were signs that the stalemate between Qatar and Saudi Arabia might be over. See Fiona MacDonald, "Signs Mount That Qatar Embargo May End as Mediation Gears Up," Bloomberg, November 13, 2019, https://www.bloomberg.com/news/articles/2019-11-12/saudi-uae-to-attend-gulf-cup-in-qatar-as-feud-begins-to-fade (accessed April 20, 2020).

40. Dominic Dudley, "Is Time Running Out for the Gulf Cooperation Council?," *Forbes,* July 25, 2018, https://www.forbes.com/sites/dominicdudley/2018/07/25/is-time-running-out-for-the-gulf-cooperation-council/#5160bea356b8 (accessed April 18, 2020).

41. Among the top 20 oil producing countries of the world https://en.wikipedia.org/wiki/List_of_countries_by_oil_production (accessed April 5, 2020), Brazil, Venezuela, and Nigeria are equatorial, hence only mildly sensitive to climate change; Iran, Mexico, and Angola are subtropical and can likewise expect only moderate impacts as a result of global warming. Climate predictions for Iraq, Algeria, and Libya, producers who also happen to be predominantly desert countries, are less extreme than models for the Gulf. The US, Russia, China, Canada, Norway, and to some extent Kazakhstan are major oil producers where substantial territories actually stand to gain from global warming.

42. Jared Diamond's *Collapse* (New York: Penguin, 2011) vividly illustrated how past leaders who could extract and save themselves from imminent calamity betrayed their own communities at times when they needed them most.

43. Kelly Mcevers, "Saudi Arabia Tries to Stall Global Emissions Limits," NPR, December 10, 2009, https://www.npr.org/templates/story/story.php?storyId=121282056 (Accessed June 14, 2019).

44. https://en.wikipedia.org/wiki/William_C._Durant#Biography (accessed April 5, 2020).

45. Ibid.

46. B. Folsom, "Billy Durant and the Founding of General Motors," Mackinac Centre for Public Policy, 1998, https://www.mackinac.org/article.aspx?ID=651 (accessed April 5, 2020).

47. https://en.wikipedia.org/wiki/Studebaker (accessed April 5, 2020).

48. Charles Komanoff and Christopher Ketcham, "What the Coronavirus Pandemic Can Teach Us about the Climate Emergency," The Intercept, April 4, 2020, https://theintercept.com/2020/04/04/coronavirus-pandemic-climate-change-co2-emissions/ (accessed April 18, 2020).

49. https://twitter.com/MichaelGerrard/status/1243213803535437824 (accessed April 18, 2020).

50. Komanoff and Ketcham, "What the Coronavirus Pandemic Can Teach Us." https://theintercept.com/2020/04/04/coronavirus-pandemic-climate-change-co2-emissions/

Lightning Source UK Ltd.
Milton Keynes UK
UKHW011313240920
370118UK00016B/318